SERIES

AUG 30 2007

Garbage and Recycling

Other Books of Related Interest:

Opposing Viewpoints Series

Conserving the Environment

Current Controversies Series

Conserving the Environment

Pollution

"Congress shall make
no law ... abridging
the freedom of speech,
or of the press."

First Amendment to the U.S. Constitution

The basic foundation of our democracy is the First Amendment guarantee of freedom of expression. The Opposing Viewpoints Series is dedicated to the concept of this basic freedom and the idea that it is more important to practice it than to enshrine it.

Garbage and Recycling

Mitchell Young, Book Editor

GREENHAVEN PRESS

An imprint of Thomson Gale, a part of The Thomson Corporation

Detroit • New York • San Francisco • New Haven, Conn. • Waterville, Maine • London

Christine Nasso, *Publisher*
Elizabeth Des Chenes, *Managing Editor*

© 2007 Thomson Gale, a part of The Thomson Corporation.

Thomson and Star logo are trademarks and Gale and Greenhaven Press are registered trademarks used herein under license.

For more information, contact:
Greenhaven Press
27500 Drake Rd.
Farmington Hills, MI 48331-3535
Or you can visit our Internet site at http://www.gale.com

Cover photograph reproduced by permission of freephotos.com

LIBRARY OF CONGRESS CATALOGING-IN-PUBLICATION DATA

Garbage and recycling / Mitchell Young, book editor.
  p. cm. -- (Opposing viewpoints)
  Includes bibliographical references and index.
  ISBN-13: 978-0-7377-3651-9 (hardcover)
  ISBN-13: 978-0-7377-3652-6 (pbk.)
  1. Refuse and refuse disposal. 2. Recycling (Waste, etc.) I. Young, Mitchell.
  TD791.G37 2007
  363.72'8--dc22
                                                                    2007004374

ISBN-10: 0-7377-3651-8 (hardcover)
ISBN-10: 0-7377-3652-6 (pbk.)

Printed in the United States of America
10 9 8 7 6 5 4 3 2 1

Contents

Chapter 1: How Do Political and Social Systems Affect Garbage Disposal?

Chapter 2: Is Recycling Environmentally and Economically Successful?

Chapter 3: Do Specific Types of Waste Pose a Threat?

Chapter 4: Can New Technologies Solve Waste Problems?

Why Consider Opposing Viewpoints?

> *"The only way in which a human being can make some approach to knowing the whole of a subject is by hearing what can be said about it by persons of every variety of opinion and studying all modes in which it can be looked at by every character of mind. No wise man ever acquired his wisdom in any mode but this."*
>
> John Stuart Mill

In our media-intensive culture it is not difficult to find differing opinions. Thousands of newspapers and magazines and dozens of radio and television talk shows resound with differing points of view. The difficulty lies in deciding which opinion to agree with and which "experts" seem the most credible. The more inundated we become with differing opinions and claims, the more essential it is to hone critical reading and thinking skills to evaluate these ideas. Opposing Viewpoints books address this problem directly by presenting stimulating debates that can be used to enhance and teach these skills. The varied opinions contained in each book examine many different aspects of a single issue. While examining these conveniently edited opposing views, readers can develop critical thinking skills such as the ability to compare and contrast authors' credibility, facts, argumentation styles, use of persuasive techniques, and other stylistic tools. In short, the Opposing Viewpoints series is an ideal way to attain the higher-level thinking and reading skills so essential in a culture of diverse and contradictory opinions.

In addition to providing a tool for critical thinking, Opposing Viewpoints books challenge readers to question their own strongly held opinions and assumptions. Most people form their opinions on the basis of upbringing, peer pressure, and personal, cultural, or professional bias. By reading carefully balanced opposing views, readers must directly confront new ideas as well as the opinions of those with whom they disagree. This is not to simplistically argue that everyone who reads opposing views will—or should—change his or her opinion. Instead, the series enhances readers' understanding of their own views by encouraging confrontation with opposing ideas. Careful examination of others' views can lead to the readers' understanding of the logical inconsistencies in their own opinions, perspective on why they hold an opinion, and the consideration of the possibility that their opinion requires further evaluation.

Evaluating Other Opinions

To ensure that this type of examination occurs, Opposing Viewpoints books present all types of opinions. Prominent spokespeople on different sides of each issue as well as well-known professionals from many disciplines challenge the reader. An additional goal of the series is to provide a forum for other, less known, or even unpopular viewpoints. The opinion of an ordinary person who has had to make the decision to cut off life support from a terminally ill relative, for example, may be just as valuable and provide just as much insight as a medical ethicist's professional opinion. The editors have two additional purposes in including these less-known views. One, the editors encourage readers to respect others' opinions—even when not enhanced by professional credibility. It is only by reading or listening to and objectively evaluating others' ideas that one can determine whether they are worthy of consideration. Two, the inclusion of such viewpoints encourages the important critical thinking skill of objectively evaluating an author's credentials and bias. This evaluation will illuminate an author's reasons for taking a particular stance on an issue and will aid in readers' evaluation of the author's ideas.

It is our hope that these books will give readers a deeper understanding of the issues debated and an appreciation of the complexity of even seemingly simple issues when good and honest people disagree. This awareness is particularly important in a democratic society such as ours in which people enter into public debate to determine the common good. Those with whom one disagrees should not be regarded as enemies but rather as people whose views deserve careful examination and may shed light on one's own.

Thomas Jefferson once said that "difference of opinion leads to inquiry, and inquiry to truth." Jefferson, a broadly educated man, argued that "if a nation expects to be ignorant and free . . . it expects what never was and never will be." As individuals and as a nation, it is imperative that we consider the opinions of others and examine them with skill and discernment. The Opposing Viewpoints series is intended to help readers achieve this goal.

David L. Bender and Bruno Leone,
Founders

Introduction

Twenty years ago, only one curbside re-cycling program existed in the United States, which collected several materials at the curb. By 2005, almost 9000 curb-side programs had sprouted up across the nation. As of 2005, about 500 mate-rials recovery facilities had been estab-lished to process the collected materials.

—U.S. Environmental
Protection Agency, 2006

In the late 1980s, garbage was on America's collective mind. Stories in the media, among them the tale of the wandering garbage barge *Mobro 4000*, provoked concern about a shortage of places to dump the country's refuse. A quote from a 1988 *Time* magazine story captures the public mood: "I'm horrified by the fact that we generate so much garbage and don't have a place to put it," exclaimed a visitor to Chicago after seeing refuse dumped in the streets.

George J. Church, the author of the above-quoted article wrote that this was a "simple but accurate description of a situation approaching the crisis stage throughout the United States. The affluent, fast-paced, throwaway American culture is producing trash on a stupendous scale." Existing municipal landfills were filling up rapidly. Objections from residents near proposed dump sites prevented construction of new facilities. At the time this attitude was derided as NIMBY-ism (for Not In My BackYard), yet there were real concerns with the smell and traffic generated by municipal landfills. More importantly, the dumps were polluting groundwater, leading to serious health issues.

Fast-forward two decades and it seems as though the "garbage crisis" has passed. Landfill capacity in the United States has increased, even while the number of landfills have decreased. Recycling—despite a large amount of skepticism on the part of commentators—has made some impact on the amount of waste that Americans generate, particularly in the composting of yard clippings. The impression that we no longer need be concerned with the waste generated by our affluent society may be misleading, however. The development of so-called mega-landfills has made it possible to deposit a large amount of waste far from the cities and suburbs where the waste is generated, but environmental activists claim these facilities encourage the production of ever-more garbage and may not be environmentally safe. Environmental justice advocates claim that much of the burden of storing our waste has been given to poor, minority, and rural communities, which often agree to the siting of garbage-processing facilities in the hopes that they will create jobs. Rather than solving the problem, then, have we in fact dumped our refuse problem on the poor and politically less powerful?

Environmental activists would answer yes, the burden definitely has been shifted to more rural, poorer areas of the country. Some localities welcomed the garbage as a source of revenue. In an article by David Taylor in *Environmental Health Perspectives*, an official of King and Queen County, Virginia, was quoted as saying: "We built the new courthouse, a new administration building, and have been able to increase the budget for our schools. It's just been a big, big blessing," This view was shortsighted, however, according to the director of the Virginia branch of the Sierra Club. "The question is, who's going to clean up a megafill when it leaks 30 years down the line? Citizens of Virginia, not the county, are going to be stuck with the bill."

Virginia eventually had second thoughts about importing other states' garbage; in 1999 it sought to ban the practice, but the ban was overturned because it was contrary to the Constitution's prohibition on states' regulating interstate commerce. Virginia has had more luck enforcing strict regulations on garbage processors; for example passing legislation to prevent the leakage of "garbage juice" into the state's rivers by mandating better construction of garbage barges. The battle continues, as other rural states rethink the policy of hosting mega-landfills. For example, in 2006, North Carolina passed a moratorium on the construction of huge waste facilities in order to allow more time to study their environmental, health, and economic impacts.

In Virginia and North Carolina, dumps were affecting a rural, largely white population that tended to be poorer than the national average. Environmental justice advocates point out that dumping practices also affect minorities disproportionately. From 1996 to 1998 the town of Chester, Pennsylvania, was a prominent case at the center of the battle over race and waste disposal. Residents filed a lawsuit alleging that the state had deliberately sited an incinerator and other waste disposal facilities in the town precisely because of its racial demographics; 65 percent of its population is African American. According to the activists, the town suffered from maladies including an increased incidence of lung cancer and a high proportion of low-birth-weight babies. Using civil rights law, the activists were able to stop the permitting process for more waste disposal facilities in the area. There have been other successes in using the concept of environmental racism to halt the building of waste treatment facilities. Citizens in Macon County, Georgia, for example, rallied to stop development of a mega-landfill near the famous Tuskegee Institute, a historically black college.

Waste processors, federal agencies, and even some state and local governments believe that efforts to stop mega-landfills and other garbage treatment facilities, like the campaign in Chester, are counterproductive. They note that modern landfills are much safer than pre-1990s municipal landfills. According to the Environmental Protection Agency, new technologies like "bioreactors"—which break down waste and reduce its volume—lower waste toxicity and mobility. Waste facilities provide jobs and revenue, which continue to be an incentive for rural communities. In late 2006, citizens in Camden County, North Carolina, expressed support for building a mega-landfill in their area. "Up to 100 jobs could be created. We would have money for schools and county infrastructure. It would help the business park that should be in front of it," said one resident testifying before the county commissioners.

It is difficult to see how those who view landfills as dangerous and those who see them as a source of jobs can ever agree. No doubt controversies over garbage and recycling will continue over the coming years. Population and thus waste generation are increasing. New forms of waste, such as that from obsolete electronic devices, pose new challenges. These will lead to further political battles over how and where waste is dumped. These issues are explored in further detail in *Opposing Viewpoints: Garbage and Recycling* in the following chapters: How Do Political and Social Systems Affect Garbage Disposal? Is Recycling Environmentally and Economically Successful? Do Specific Types of Waste Pose a Threat? Can New Technologies Solve Waste Problems?

CHAPTER 1

How Do Political and Social Systems Affect Garbage Disposal?

Chapter Preface

A round 200 B.C. the Greek city-state of Athens designated an area outside its walls for the dumping of garbage. This is the first recorded example of using the political process to manage the waste disposal issue. In our own time, politics also plays a large role in how we dispose of our garbage. Our social attitudes determine, in part, how much garbage we generate. How to dispose of refuse continues to be a political and social question in addition to a technological challenge.

A major division in attitudes toward the disposal of garbage often involves a clash of views about the proper degree of government involvement in our lives. Partisans of individualism and the free market generally take the view that the United States has plenty of land available for garbage disposal. We can afford to generate and bury vast amounts of garbage, they argue. The public should not be concerned about being buried in trash, nor should politicians use a dubious garbage crisis to mandate sharp reductions in the amount of waste put into landfills, they maintain. "Despite the 'garbage trucks could ring the Milky Way galaxy' rhetoric, all of the trash America will produce over the next 1,000 years could fit into a landfill 15 square miles in size," writes libertarian columnist Doug Bandow. As a believer in strictly limited government, Bandow exemplifies the view that we can continue to give little thought to garbage. If, one day, land prices or the value of the metal, plastic, or glass in goods increases due to scarcity, consumers will automatically reduce consumption and increase efforts at reducing waste.

Environmentalist thinkers, in contrast, argue that this view is shortsighted because the amount of garbage per person in the United States continues to increase despite recycling efforts. Some believe that the American lifestyle is to blame.

"The simple answer is that we buy too much: too much food we don't eat, too many clothes we hardly wear, too many quickly obsolete electronic gadgets, and way too much packaging (almost a third of the average city's trash)," writes Jennifer Hattam of the environmental group the Sierra Club. Environmentalists argue that even if we have land available today for landfills, we must worry about future generations; we cannot continue to produce a vast quantity of waste per person without causing problems for our children and our grandchildren.

The following articles discuss issues of waste generation and waste disposal from social and political perspectives, providing a variety of viewpoints on the question of how waste is generated, who is most impacted by waste, and how best to minimize that impact.

"The most immediate role for senior governments is to establish the policy and regulatory framework within which [waste] producers will take responsibility for the environmental management of their products."

Current Municipal Waste Systems Lead to Dangerous Landfills

Helen Spiegelman and Bill Sheehan

In the following viewpoint, Helen Spiegelman and Bill Sheehan trace the history of the increase in solid waste. Noting the growth of packaging and old products in the waste stream, Spiegelman and Sheehan propose the passage of laws requiring manufacturers to pay for the safe recycling—"cradle to cradle"—of the goods they produce, thus ending the "free ride" that the current municipal waste collection system provides to companies.

Helen Spiegelman is the president of the Product Policy Institute, an environmental research and policy organization, and Bill Sheehan is the institute's secretary.

Helen Spiegelman and Bill Sheehan, "Getting Rid of the Throwaway Society," *In Business*, vol. 28, issue 2, March/April 2006. Reproduced by permission.

As you read, consider the following questions:

1. Around 1910 what made up the vast majority of waste generated by people in New York City, according to the authors?

2. Today, what is the largest category of municipal solid waste, in Spiegelman and Sheehan's opinion?

3. According to the authors, who should take ultimate responsibility for the environmental management of products?

The earliest municipal waste managers characterized municipal refuse using three categories: ashes, garbage and rubbish. Ashes were the residue of coal and wood used for space heating and cooking; garbage was the putrescible [biodegradable] wastes produced in food preparation; and rubbish was a miscellaneous category made up of various worn out products and packaging. Surveys conducted during the early part of the 20th century found that ashes were by far the largest category of refuse. "Between 1900 and 1920," writes historian Martin Melosi, "each citizen of Manhattan, Brooklyn, and the Bronx annually produced about 160 pounds of garbage, 1,231 pounds of ashes, and 97 pounds of rubbish." According to Melosi, these figures fall pretty much within the ranges identified in other studies where garbage ranged from 100–180 pounds, ashes from 300–1,000 pounds, and rubbish from 50–100 pounds.

In addition to the wastes collected from households and businesses, municipal waste managers faced a staggering quantity of organic wastes generated by the horses that served as the main means of transport in cities. Each horse was estimated to produce 20 pounds of manure and gallons of urine daily. In addition, since city horses had a life expectancy of only a couple years, their carcasses were an additional burden. Melosi wrote in his book, *Garbage in the Cities*, that as late as

1912, when motor vehicles already dominated the streets, scavengers removed as many as 10,000 horse carcasses from the streets of Chicago.

Not surprisingly, municipal refuse was seen as an urgent public health problem. It only made sense to provide for the prompt removal of putrescible waste as a community service. This was one of the Progressive era reforms instituted to make life more bearable in the growing industrial cities of North America. Over time, the entrepreneurial rag and bone man was put out of business by uniformed municipal crews that hauled the community's refuse to an official disposal site.

Waste As an Environmental Problem

By the 1960s municipal solid waste was beginning to be viewed as an environmental problem as well as a threat to public health. Groundwater impacts from landfills and air pollution from waste incinerators were a continuing concern, but there was also a growing policy emphasis around resource conservation and materials recycling. In 1969 the National Environmental Policy Act made a commitment for the federal government to, among other things, "enhance the quality of renewable resources and approach the maximum attainable recycling of depletable resources."

In pursuit of this policy, the U.S. Environmental Protection Agency (US EPA) produced annual reports characterizing the municipal solid waste in the United States. The reports divide municipal waste into two basic categories: products and other wastes. "Products" are manufactured goods and packaging—what was earlier called "rubbish." "Other wastes" are primarily food scraps and yard trimmings (called "garbage" in 1900) plus a now small amount of inorganic wastes (which would include the ashes in earlier times). This typology allows easy comparison with the waste surveys from the beginning of the 20th century.

Most obvious are the disappearance of ashes (today classified as an industrial rather than a municipal waste) and the growth of product-related waste. Note also that a new category of nonproduct waste emerged in the 20th century—yard trimmings—reflecting the suburbanization of North America. We refer to the combined food scraps and yard trimmings as "biowastes."

Insignificant a century ago, products are now the largest category of municipal solid waste by far, comprising fully three-quarters of total waste by weight in 2000. Between 1960 and 2000, product waste generation more than tripled in total tonnage, from 54.6 to 176.3 million tons per year, while the generation of biowastes grew more slowly, at about the rate of population growth, from 33.5 to 57.7 million tons per year. . . .

Product Discards Drive Waste Growth

The main driver in the growth of waste throughout the 20th Century has been product wastes. Despite efforts in recent years, our municipal solid waste management system has been unable to have an impact on consumption, which is driven by factors beyond community control.

In fact, the provision of universal collection and disposal of waste creates conditions that actually encourage the production and consumption of throwaway products. Throughout the 20th century, local communities have provided convenient removal of any product that the householder had no more use for. This has acted as a perverse subsidy, encouraging production and consumption of short-lived products. In this way, cities and towns have been enablers of our growing addiction to convenience, facilitating the excessive material flows that characterize our consumer society and put humans in conflict with the environment. Municipal waste management has also been a form of corporate welfare for the companies that sell throwaway products, allowing them to win customers by promising "convenience" that is provided at public expense.

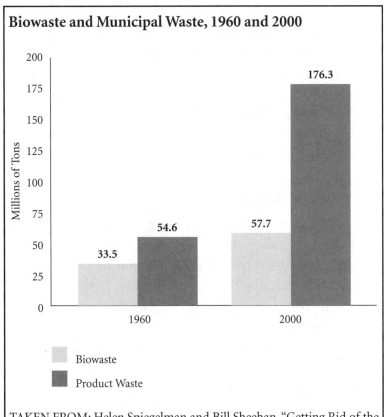

Biowaste and Municipal Waste, 1960 and 2000

TAKEN FROM: Helen Spiegelman and Bill Sheehan, "Getting Rid of the Throwaway Society," *In Business*, March/April 2006.

The municipal solid waste management system, originally configured to manage wastes made up of relatively homogeneous materials such as ash and biowastes, is entirely unsuited for managing today's complex consumer products. As a result, products with toxic components as well as those with valuable, highly engineered materials are collected in packer trucks and deposited en masse at landfills or waste incinerators. Even in municipal recycling programs, bulk handling degrades the value of the collected materials, with the result that as much as half of some collected materials, such as glass, ends up in landfills or low-value applications rather than replacing demand for raw materials.

Watching their recycling volumes level off and their overall waste generation continue to climb, municipal waste managers have started to shift their focus away from recycling. US EPA reports that the number of curbside recycling programs declined from 9,700 to 8,875 between 2001 and 2002. The municipal waste management system is effectively conceding defeat in the goals established in the 1980s to stem material flows and conserve resources.

An Alternative Approach

The Product Policy Institute sees a clear message in these data: municipal waste policy should distinguish between product wastes and biowastes. Product wastes should be managed in a separate system, described below, rather than in the municipal waste system. The municipal waste management system should be focused around programs that provide biowaste management. This could be either through public facilities owned and operated by the community or through facilities owned and operated by the private sector or some combination of the two. Public policy should encourage organics processing facilities to emerge, scaled to serve local markets (minimizing hauling) and numerous enough to provide stability to the overall system. Public policy should encourage a range of technologies serving different commercial clients such as grocery retailers, restaurants, etc., as well as community residents and landscapes.

Product recycling, on the other hand, should be seen as an extension of the product marketing system, rather than an extension of the municipal waste management system. The recovery and recycling of a product should be managed through commercial arrangements made between the product brand owner and supply chain contractors. In this way, the product recycling process will mirror the production and distribution process.

Policy Support for Waste Management

The most immediate role for senior governments is to establish the policy and regulatory framework within which producers will take responsibility for the environmental management of their products. The legislative model developed in British Columbia (Canada) has been effective, providing the opportunity for brand owners to develop a program that suits their needs, while at the same time establishing clear environmental benchmarks and provisions for enforcement to prevent "free-riders" and protect the environment and public health and safety.

The regulatory process serves everyone best if regulations are harmonized, reflecting the highest common environmental and social standards as well as respecting the needs of companies that market products in a global marketplace. The European Union establishes "directives" that are implemented with legislation at the national level. Such a model might apply to trading partners within the North American Free Trade Agreement.

At the local level, biowaste processing should be supported with transitional policy instruments such as differential disposal fees or outright bans on disposal, while product recycling services are supported through zoning and business licensing tools.

By separating the functions of product and biowaste management, this approach encourages innovation by producers to avoid waste, and also provides for the safe and beneficial use of community biowastes.

*"In order to handle all the waste pro-
duction of the twenty-first century, each
state merely needs to find space for a
single, square landfill, 2.5 miles on each
side."*

The United States
Has Room for
Twenty-First-Century Garbage

Bjørn Lomborg

Bjørn Lomborg's book The Skeptical Environmentalist *was quite
a shock to the environmental community. The Danish professor
of statistics held that much of what he called "the litany" of environmental complaints was exaggerated. In this viewpoint taken
from the book, Lomborg claims that America has plenty of room
for landfills. He forecasts the amount of solid waste that the
United States will generate over the course of the twenty-first
century and shows that landfills will require a very small percentage of the total land mass of the country.*

As you read, consider the following questions:

1. How much solid waste does each American on average
   produce per day, according to the author?

Bjørn Lomborg, from *The Skeptical Environmentalist: Measuring the Real State of the
World*. New York: Cambridge University Press, 2001, pp. 206–209. Reproduced with
the permission of Cambridge University Press.

2. How many square miles of landfill does Lomborg say will be required for U.S. waste in the twenty-first century, assuming the waste is piled one hundred feet high?

3. Which nation has the highest per capita waste generation: Japan, Germany, or France, according to the author?

We often worry about all the waste piling up, wondering where it all can go. We feel that the "throwaway society" and its industrial foundation is undermining the environment. This fear is perhaps expressed most clearly by former vice president Al Gore, who is disturbed by "the floodtide of garbage spilling out of our cities and factories." "As landfills overflow, incinerators foul the air, and neighboring communities and states attempt to dump their overflow problems on us," we are now finally realizing that we are "running out of ways to dispose of our waste in a manner that keeps it out of either sight or mind." The problem is that we have assumed "there would always be a hole wide enough and deep enough to take care of all our trash. But like so many other assumptions about the earth's infinite capacity to absorb the impact of human civilization, this one too was wrong." Equally, Isaac Asimov in his environmental book tells us that "almost all the existing landfills are reaching their maximum capacity, and we are running out of places to put new ones."

Waste Crisis Fails to Materialize

It is true that waste generation does increase with GDP [gross domestic product]. The richer we get the more garbage we produce.... The question is, of course, whether this is actually a problem. We may believe that garbage production is spiraling out of control and that landfill garbage is piling up to the extent that soon there will be room for no more, but that is simply not the case.

A waste expert [Marian Chertow] points out that reality has turned out very differently from our fears of just a decade ago: "Images from the nightly news of more and more garbage with no place to go struck fear in the hearts of mayors and public works directors everywhere. Children were taught that the best way to stave off the invasion of sea gulls hovering at landfills was by washing out bottles and stacking old newspapers. But the anticipated crisis did not occur."

Each American produces about 4.5 pounds of waste per person each day—all in all some 200 million tons of municipal waste each year. Not only does that sound like a lot, but the annual amount has doubled since 1966. . . . However, the growth in waste that actually ends up in the landfill has stopped growing since the 1980s, and currently Americans ship off less waste to the landfill than they did in 1979. The main reason is that more and more waste gets incinerated, recycled or composted. Moreover, a part of the reason why the US produces more waste is more people—per person, total waste has only increased 45 percent since 1966. And looking at the waste that will end up in a landfill, each person only produces 13 percent more waste than in 1966.

EPA's [Environmental Protection Agency] data only go back to 1960 but consumption patterns have changed radically over longer periods. At the beginning of the twentieth century, an American home each day produced in addition to the ordinary waste some 4 pounds of coal ash, making it likely that landfill production has not increased dramatically over the century.

Landfills Will Require Little Space

Nevertheless, it seems likely Americans will continue producing at least 110 million tons of garbage every year destined for the landfill. Natural intuition, as pointed out by Al Gore above, tells us that this cannot go on forever. If we envisage that the US will continue to produce 110 million tons of landfill waste every year for the rest of this century, from 2001 to 2100, how much space would it occupy? Let us suppose we placed all the waste in a single landfill—surely not a smart idea, but just for the purpose of illustration—and filled it up to 100 feet [high]. That height is still lower than the Fresh Kills landfill on Staten Island, within the New York City boundaries. Then the total landfill waste of the US over the entire century would take up just a square 14 miles on each side.

Now, surely, assuming a stable waste production for the next century is an unreasonably optimistic assumption. Not only will economic growth increase the amount of garbage, but according to the Census Bureau, the US is expected to more than double its population before 2100. Thus, let us assume the growth in total personal waste production as we have seen since 1990 and expect till 2005 will continue till 2100. Moreover, let us adjust the waste production to the ever increasing number of Americans, each producing ever more garbage. Once again, sum up the entire waste mass and pile it to 100 feet. Surprisingly, we will only need a slightly larger area—it will still fit within a square, less than 18 miles on each side.

... All the American waste of the entire twenty-first century will fit into a single landfill, using just 26 percent of the Woodward County [Oklahoma] area. Of the state of Oklahoma, the landfill would use up less than half a percent. Of the entire US landmass, the landfill would take up about one-12,000th—less than 0.009 percent. Equivalently, one could imagine that each state would handle its own garbage—for argument's sake, let us just say one-fiftieth each. Then, in order to handle all the waste production of the twenty-first century, each state merely needs to find space for a single, square landfill, 2.5 miles on each side.

Moreover, the scenario with ever increasing amounts of waste is probably rather exaggerated, especially considering that most economic growth will be in the service industries and information technology.... Even in material production the general trend is towards the use of fewer materials—a sort

of dematerialization of the economy. The car is an excellent example, representing a full basket of products from an industrialized economy, with metals, plastics, electronic materials, rubber, and glass. Since the early 1970s, carbon steel has been replaced with high-tech steel, plastics and composites, with the new materials substituting the old at a rate of one to three, making the car ever lighter without compromising structural integrity.

Even so, the main point here is that we will not be inundated with garbage. Garbage is something we can deal with. It is a management problem.

This does not, however, imply that landfills will be easy to site. Nobody wants to be neighbor to a landfill—a phenomenon so familiar that it has even been given a name: NIMBY, or Not In My Back Yard. Thus, garbage may be a *political* problem, but it is not a problem of lack of physical space.

It is perhaps worth mentioning that landfills today are very safe for the groundwater. The EPA estimates that current environmental regulations governing the US's 6,000 landfills ensure that over the next 300 years they will only cause 5.7 cancer-related deaths, or just one every 50 years. This should be seen in the light of the fact that cancer kills 563,000 people each year in the US, with about 2,000 deaths caused merely by using spices in food.

For other nations, waste rates seem to be slightly increasing but at much lower levels than the US. At 1.1 kg/day [kilogram/day] for Japan and 1.3 kg/day for France, both have seen slow increases, but they are still way below the American production of 2 kg/day. As a result of strict policies, German waste production of 1.2 kg/day has actually decreased 29 percent since 1980. The UK has very poor waste statistics, but seemingly the trend is slightly upwards, with a daily production on a par with France. If UK waste production increases at the same rate as the American (surely an overestimate, since the UK population does not increase with nearly the same speed), the total landfill area needed for the twenty-first-century UK waste would be a square 8 miles on a side—an area equivalent to 28 percent of the Isle of Man.

Finally, we should mention recycling. In the US paper, glass, metals and plastics are recovered. We tend to believe that recycling is a rather new phenomenon, but actually the US has recycled about 20–30 percent of all paper throughout the century, and recycling is still below the levels of the 1930s and 1940s. However, materials like copper and lead have increasingly been reused, rising steadily through the century from 5–10 percent to more than 50 percent and 70 percent respectively.

Moreover, we tend to believe that all recycling is good, both because it saves resources and because it avoids waste. Of course, ... we may not necessarily need to worry so much about raw materials, especially common ones such as stone, sand and gravel, but neither should we worry about wood and paper, because both are renewable resources.

Equally, if the entire US twenty-first-century waste can be contained within a single landfill in part of Woodward County, Oklahoma, we must also consider whether recycling to avoid waste is a good investment of resources. Possibly, we may be able to save more resources by burning old paper at incineration plants, making use of the heat produced and felling more trees, instead of using energy to collect the old paper to be sorted, prepared and filtered. New studies seem to indicate that it actually costs more to recycle paper than to produce new paper.

Societal-based analyses typically show that recycling does not pay from a private economic point of view, although it is in the balance as far as society as a whole is concerned. This can be seen as evidence that the current recycling level is reasonable, but that we perhaps should not aim to recycle much more.

> *"In their own way, these comprehensive [recycling] programs address the significance of individual choice in the creation of waste."*

Mandatory Recycling Promotes Environmental Awareness

Heather Rogers

Heather Rogers is a journalist and documentary filmmaker. In this viewpoint she argues that laws that force manufacturers to "internalize"—pay for—the cost of disposing the packaging in which they sell their products show that waste reduction efforts can be successful. In Germany, which has such a law, excess packaging has decreased and recycling has surged. Such laws, in effect not only in Germany but in some American cities, have an important benefit in that they make consumers aware of the environment and create a culture of recycling where they are instituted.

As you read, consider the following questions:

1. What type of garbage is covered by Germany's "Green Dot" program, according to the author?

Heather Rogers, *Gone Tomorrow: The Hidden Life of Garbage*. The New Press, 2005. © 2005 by Heather Rogers. Reproduced by permission of The New Press, www.the newpress.com. (800) 233-4830.

2. Under the "extended producer responsibility" concept, who should pay for the disposal of waste such as packaging or obsolete electronic equipment, according to Rogers?

3. What American city mentioned by the author has a program of sorting garbage similar to Germany's?

In a striking move, the conservative German government of Helmut Kohl passed the landmark 1991 Packaging Ordinance, a law that shifted the burden of collecting, sorting, recycling and disposing of packaging wastes away from taxpayers and onto manufacturers. A version of "extended producer responsibility," the ordinance forced industry to administer and bankroll a separate refuse handling system just for packaging. Amid fierce resistance from business and cheers from environmentalists, the Packaging Ordinance marked a serious intervention by government in the decisions of production, disallowing industry's full externalization of commodity waste costs. Not without its limits and flaws, Germany's still-active law has served as a model program for addressing the single largest component of the discard stream—packaging.

Producers Pay for Costs of Waste Disposal

The mandatory nationwide ordinance allowed the manufacturing sector to formulate its own methods for handling trashed wrappers and containers; the result was what came to be known as the "green dot" system. Still in operation, the program is overseen by the environmental ministry and administered by an industry-led entity called Duales System Deutschland (DSD). The procedure is simple: After consumers make their purchases, they can discard the wrapping at the store, or, once home, they can place all their spent packaging into a yellow bin or bag that gets collected at the curbside by DSD.

DSD financing comes from the country's private producers, who are required to pay a licensing fee proportionate to the cost of handling their specific packaging material based on weight and type. To signal payment, these manufacturers get to stamp their cellophane, boxes, bottles and cans with the green dot symbol—two arrows swirling together—an image that not only helps consumers sort their discards, but also, like the recycling symbol in the United States, generates excellent eco-PR.

Germany's program was the first successful large-scale public implementation of extended producer responsibility (EPR). A new spin on waste reduction strategies proposed in the 1970s, EPR has gained popularity in the past decade and aims to shift from the public to manufacturers the duties of managing discarded commodities like packaging and the mushrooming piles of e-waste. The underlying idea is to persuade companies to generate less from the start. If they have to pay to handle and treat these wastes, the logic goes, producers will ultimately choose to create fewer disposables.

Increased Recycling

According to economist Frank Ackerman, the 1991 ordinance "jump-started producer responsibility—it was proof of possibility." Ackerman also explains that in the face of the German program's indisputable success, government's active role in curtailing waste production could no longer be discredited: "It was such a striking example of going from nothing to accomplishing this complex task. The government decided to do it and did it, proving that producer responsibility could actually be implemented."

The green dot system invalidated opponents' predictions of economic cataclysm and claims that the state could not successfully regulate industry for environmental protection. Hailed a success, the program significantly cut packaging consumption, boosted recycling, and sustained the use of refillable bottles. In the ordinance's first half-decade, manufacturers used 7 percent less packaging; by contrast during that same period, U.S. container consumption grew by 13 percent. Obligatory recycling targets under the Packaging Ordinance, which increased incrementally over time, have kept DSD on track. As a result, recycling for packaging has surged by 65 percent since 1991. And the ordinance's requirement that at least 72 percent of the country's beverage containers be refillable has guaranteed reduced consumption by requiring reuse. The program has proven so beneficial that, since the mid-1990s, fifteen European Union countries as well as Korea, Taiwan and Japan all enacted variations on Germany's Packaging Ordinance.

Program Weaknesses

Unfortunately, the German system has its downsides. In more recent years packaging production in the country has increased, despite the green dot program. As of 2003, Germany was the largest single market for packaging throughout Europe. So, while producers may be financing the handling of their packaging wastes, they are not deterred from producing these discards in ever-greater amounts. This reveals the system's limits; mere encouragement of rubbish reduction is not enough. While Germany's system forces manufacturers to internalize some refuse treatment fees, more accurately distributing production costs, it has not brought changes that directly cut waste.

Among its other shortcomings, the green dot signals only that a producer has paid its fees to DSD, guaranteeing neither that a package was made from recycled materials nor that it will be reprocessed. What's more, recycling is defined so broadly under the green dot system that it includes the channeling of used plastics to the steel, petrochemical and oil industries for burning as "fuel." And since recycling in Germany is just as susceptible to market fluctuations as in the United States, materials that get collected by DSD might easily end up incinerated or landfilled instead.

German consumers sort packages bearing this symbol from other trash for recycling or disposal at the manufacturer's expense.

Creating Recycling Culture

Despite these weaknesses, the daily sifting of one's discards contributes to more widespread popular consciousness of the value of trashed materials. The German green dot is but one element in a nationwide household trash-sorting culture. In addition to segregating most packaging wastes, residents separate recyclable paper, glass and metal, while they store food castoffs in special brown bins. Consumers return empty milk and beer bottles to the supermarket and whatever is left over (not much) gets deposited into a black refuse can. In the United States, such setups are not entirely unheard of. San Francisco's recently revamped municipal system calls for extensive household segregation of castoffs—sending bottles, cans and paper to recycling centers and food scraps to a compost facility, with the much reduced remainder going to the landfill. San Francisco's new program is a sharp improvement on previous methods and, like the German system, it fosters an awareness of discards as useful, not just spent and dirty waste.

In their own way, these comprehensive programs address the significance of individual choice in the creation of waste. While the underlying structures that enforce garbage are paramount, individual practices are not entirely irrelevant. In the last century the majority of Americans have gleefully accepted, even embraced, a high-waste lifestyle. And in some ways it's understandable. There are real conveniences offered by disposable products like paper cups and diapers, while seductive packaging and entrancing new commodities are hard to resist for many shoppers. There's no doubt that bulk bins, canvas shopping totes and used plastic bags drying in the dish rack can seem like part of a more pinched, less spontaneous lifestyle. Consuming sexy packaging and the latest styles is, for large parts of the population, an intensely fun and gratifying experience. And, while people might feel bad about throwing things away, they may also get a type of pleasure from tossing out the old and bringing in the new.

These realities, constantly reinforced in mainstream society, can present cultural and political obstacles to change. Creating an atmosphere where expanded reforms protecting human and environmental safety and health seem normal and even hip—witness programs like those in Germany and San Francisco—can foster a shift in thinking about consumption that counters the dominant line. Recycling unleashes various forms of imaginings; rather than just the misperception that it actually works as a long-term solution, recycling—crucially— also opens the political and cultural imagination to other creative possibilities.

> *"It isn't likely that we'll become truly efficient about resource recovery until we've exhausted all our raw materials."*

Only Necessity Will End Wasteful Habits

Elizabeth Royte

In this viewpoint taken from Garbage Land: On the Secret Trail of Trash, *award-winning science and environmental writer Elizabeth Royte discusses consumers' attitudes toward recycling. Most people claim they are interested in environmentally friendly packaging, but companies say consumers are not willing to pay a premium for "green" products. Royte fears that people will not recycle or reduce waste voluntarily. Rather, they will reduce waste only when shortages of materials force consumers and companies to use recycled products.*

As you read, consider the following questions:

1. According to the majority of producers of food, personal care, and pharmaceutical products, as cited by the author, will consumers pay high prices for packaging made from recycled products?

2. What does Royte say that philosopher Wendell Berry meant when he wrote, "We cannot 'afford' to take care of things"?

3. What does the author believe will promote recycling?

Despite the many thousands of curbside recycling programs that accept paper, paper and other packaging waste still account for between 35 and 40 percent of the household waste in North American landfills. Americans don't care enough about recycling, it seems, and packagers have all kinds of incentives for wrapping things up in paper and plastic: to prevent theft (shoving a CD in a foot-long plastic bubble down your pants is a lot trickier than secreting just the disk), to facilitate self-service, to protect products from tampering, to provide a canvas for stickers that say "New!" or "Improved!"

Packaging Waste Is Increasing

Even the righteous European Union, which has a 53 percent overall recycling rate for paper, metal, plastic, and glass, is unable to keep pace with the growing tide of packaging materials. Waste generation is linked to economic growth, said a 2004 European Environment Agency report (no surprise there), and waste is increasing as more and more food, packaged for long-haul transport and longer shelf life, moves throughout the union. (In Europe, as in the United States, hyperwrapped convenience foods are becoming more common as potential household cooks go outside the home to earn a living.) The agency also cited the rising emphasis on health and safety for a surge in food packaging. . . .

In a 2002 survey by *Packaging World* magazine, only 30 percent of the respondents, who made food, personal care, and pharmaceutical products for consumers, said that environmentally friendly packaging was "very important." When asked if consumers would pay a premium for green packaging, 61 percent said no. In the developed world, customers expect premium goods to come in premium packages. Cosmetic companies, which charge a lot of money for very small products, were the first to understand that cool packaging bespeaks cool contents.

I wasn't immune to the allure of the shiny and new, especially when it came to clothing, which I believed had some transformative power. It was the ceaseless marketing of, and the status seeking through, new possessions that I found anathema. The United States consumes far more stuff than any other developed country. According to the biologist Edward O. Wilson, if the rest of the world consumed at our levels—with existing levels of technology—we'd require the resources of four more planet Earths. (This extrapolation raises a question: before the last megafills are full, will we run out of stuff to put in them?) According to the United Nations "Agenda 21" report, "The major cause of the continued deterioration of the global environment is the unsustainable pattern of consumption and production, particularly in industrialized countries."

"We Can't Afford to Care for Things"

The flip side of consumption and production, of course, is wasting—consigning expired and unwanted goods to their fates in a landfill or incinerator. Why is there so much of it? Scholars offer various reasons. There is functional obsolescence (brought about by technical improvements); there is style obsolescence, also known as fashion; and the plain economic fact that it is often cheaper to buy something new than to repair something old. [Philosopher] Wendell Berry wrote in 1987, "Our economy is such that we cannot 'afford' to take care of things: labor is expensive, time is expensive, money is expensive, but materials—the stuff of creation—are so cheap that we cannot afford to take care of them." A lack of connection between those who make goods and those who use them contributes to the ease with which we turn our backs on our

Why Write About Garbage?

I've always wondered whether it was better, environmentally speaking, to throw a used tissue in the toilet or in the trash. And like a lot of people, I wondered where things went, and what became of them, after I threw them 'away.' So I started keeping track of my trash, quantifying it—to learn exactly what I was rejecting. Then I began traveling with my trash. As I learned how far my garbage footprint spread, I tried my utmost to leave a smaller human stain. The tissue, by the way, should go in the toilet. But don't flush till you must!

Elizabeth Royte, 2006. www.bookbrowse.com.

possessions. It is easier, for example, to throw out an ugly ceramic pitcher made in a Taiwanese factory than it is to throw out an ugly ceramic pitcher made by a well-meaning aunt or even an anonymous local craftsperson. Increasingly, handwork is not part of the equation.

The holidays are a perfect time of year to get across the source-reduction message. According to Inform, the environmental research firm, Americans produce an additional one million tons of trash per week between Thanksgiving and New Year's Day. (Waste watchers in college towns, however, note dips in garbage generation during this period and attribute it to the departure of students. Those cities' trash spikes come when students leave in late spring.) I found it hilarious that the city of Austin [Texas], which charges residents for each bag of garbage they set on the curb year round, offers a post-Christmas amnesty. "So you can throw out as much as you want and not be penalized for celebrating a Christian holiday with intense commercialism and the attendant cast-off crap," my Austin friend Spike Gillespie said to me. . . .

A Connection with Nature

It's been said that to make a dent in our garbage problems, source reduction has to acquire the rhetorical currency of recycling. One could argue that the currency of recycling isn't exactly robust, but at least it doesn't fly in the face of the holiday media messages one encounters at every turn. [The environmental organization] Inform offers loads of "greener" holiday tips, but they seem timid, dull, and rote. Give rechargeable batteries, a low-flow showerhead, a membership in an environmental organization. Yawn. Shop at thrift stores, send electronic instead of paper holiday greetings. Ho-hum. Clearly, source reduction has a PR [public relations] problem. Compared to the sirens of high-end emporiums luring us to buy, Inform is a gray-haired spinster with an admonishing finger.

If the reality of an environmental conscience (as opposed to the idea of it) isn't chic, at least it is wholesome: it speaks to a connection with the natural world. We are all, pollutocrats and composters alike, children of the universe. And nature, as everyone knows, brooks no waste. That notion gets a lot of play in liberal recycling and design circles. "Consider the cherry tree," write [William] McDonough and [Michael] Braungart [in their book *Cradle to Cradle*]. "It makes thousands of blossoms just so that another tree might germinate, take root, and grow. After falling to the ground, the blossoms return to the soil and become nutrients for the surrounding environment. Every last particle contributes in some way to the health of a thriving ecosystem." It happens on every scale: large animals die, and their carcasses feed smaller animals, fungi, and microbes, which in turn make possible the feedstuff of larger organisms. The building blocks of life cycle endlessly, just like the nature shows that deliver this message without cease.

Humans and Pack Rats

But do animals actually produce trash? Well, leaf-cutter ants carry dead workers from their underground nests, tipping them onto outdoor mortuary piles. They haul dried-out leaf fungi from garden chambers down trails to compostlike piles. (Both are exploited for nutrients by insectivores and herbivores.) Underground, kit foxes feed their young throughout the winter months. They pay the piper in the spring, when an entire season's worth of bones, fur, and feathers has to be excavated and placed on the equivalent of their curb (these scraps become food or shelter for other creatures). Prairie dogs, like humans, continually drop skin flakes and discard food in their burrows. Fleas, mites, lice, and other bottom-feeders constantly vacuum this stuff up: when the live-in maids become too populous, the original inhabitants clear out. Perhaps the animal most like humans, in its dealings with trash, is the pack rat. The desert rodents live in five-foot-high dens built of just about anything they can find (including branches, cactus pads, newspapers, cow pies, cans, and rags). When they periodically clean house or renovate, they create trash middens nearby. Varnished by coat after coat of pack rat urine, these heaps persist for ages in the dry desert. Just as William Rathje and his archaeology students analyzed landfills to suss the habits of Americans who lived a half-century earlier, scientists at the Desert Research Institute in Nevada examine the radioactive fingerprints of crystallized pack rat urine and scrutinize seeds and pine needles to learn about climate and plant communities up to 25,000 years ago.

Necessity Will Drive Recycling

Could humans learn something from nature's constant cycling of resources? Sure, but there is a crucial difference between Homo sapiens and other species. In the natural world, no organism self-sacrifices for the good of the environment. Plants and animals steal whatever resources they can, heedless of fellow creatures and the future. Nature is both terrifically efficient and terrifically wasteful, all at the same time. Green thinkers like to focus on just one aspect of this equation. It's a feel-good idea. But nature isn't wise or farsighted. Recycling, however, is wise precisely *because* it's farsighted. Unfortunately, it isn't likely that we'll become truly efficient about resource recovery until we've exhausted all our raw materials (at which point the planet will be a fairly dismal place to live). Recycling is the name we've given to resource recovery before it's profitable. Only later, when recovery becomes a profitable necessity—because all the new material is gone—will we really be living like the animals.

"Studies ... found correlation between communities of color and exposure to a variety of environmental hazards."

Racial Minorities Suffer Disproportionately from Toxic Waste

Eileen Gauna and Sheila Foster

The environmental justice movement is concerned with the disparate impact of waste and pollution on low-income and/or minority communities. In this viewpoint, two authors involved with environmental justice issues give a brief history of the movement. They then describe some of the challenges activists face in fighting the pollution that adversely impacts these communities. Law professor Eileen Gauna has written a casebook, Environmental Justice: Law, Policy and Regulation. *Sheila Foster, also a professor of law, has written* From the Ground Up: Environmental Racism and the Rise of the Environmental Justice Movement.

As you read, consider the following questions:

1. When and where was one of the earliest environmental justice events held and what were the organizers protesting, according to the authors?

2. What did the former chief economist of the World Bank say about the dumping of waste in less-developed countries, according to Gauna and Foster?

3. How do demographic shifts affect environmental justice activism, in the authors' view?

Like the civil rights movement that preceded it, the U.S. environmental justice movement was propelled into mainstream political discourse and popular consciousness, by grassroots activism long before the terms "environmental racism" or "environmental justice" were coined. Ordinary men and women were thrust into extraordinary leadership roles as they struggled against environmental degradation in their communities and the decision makers who controlled their environment. They organized efforts against businesses and institutions and environmental, land, and transportation practices that too often left them bearing a greater share of pollution than their usually whiter and more affluent neighbors.

These activists—low income and/or people of color—were fed up with living in the shadows of industrial facilities, contaminated land, transportation corridors, concentrated animal feeding operations, mining operations, contaminated aquifers, and other risk-producing and resource-depleting practices. They were also sick—literally—of unbearable smells, dust, noise, poisonous air, unexplained skin rashes, birth defects, respiratory illnesses, and rare cancers. Urban communities started to demand the right to clean, safe environments. Tribes and other indigenous peoples wanted recourse in the face of ruined sacred sites and degraded rangelands and forests. During the last twenty years, these diverse, localized grassroots efforts have coalesced into a vibrant national and international political movement, leading protests, bringing lawsuits, and influencing policy at the highest levels of government.

The History of the Environmental Justice Movement

The environmental justice movement began building momentum in the early 1980s, sparked by direct action protests over the siting of hazardous waste facilities in communities of color. The earliest of these high-profile events was a nonviolent 1982 demonstration against a polychlorinated biphenyl (PCB) landfill in predominantly African American Warren County, North Carolina. Although plans for the landfill went forward, the action led to more than 500 arrests and generated increased public interest in the idea that minority areas were targets for hazardous land use. Arguably the best-known work documenting this inequitable distribution was the United Church of Christ Commission for Racial Justice's landmark 1987 national study finding that race was the most significant variable in determining the location of risk-producing facilities.

Later studies similarly found correlation between communities of color and exposure to a variety of environmental hazards. A 1991 review performed by the Environmental Protection Agency (EPA) confirmed that racial and ethnic minorities were disproportionately exposed to air pollutants, contaminated fish, agricultural pesticides, and adverse health effects from exposure to lead, especially in children. The same year, the *National Law Journal* found significant disparities in the enforcement of federal environmental laws—for example, penalties for violations ran nearly 500 percent higher in predominantly white communities. Contaminated sites in non-white areas were targeted for cleanup more slowly, and cleanup took longer and was less protective than in predominantly white areas.

These findings sparked a controversy about the methodology of the studies. Various academics and commentators debated the underlying causes behind the disparities, posing a "chicken or egg" question about which came first—the hazardous facility (for noxious land use) or the minority community. Their premise was that poor, minority populations were "coming to the nuisance"—i.e., moving to neighborhoods that already had these hazardous facilities and, thus, cheap housing and land. Interestingly, one national study found that very poor areas tend to repel, rather than attract, hazardous waste facilities (contradicting the theory that low-cost land is the primary factor in a siting decision).

A "Chicken or Egg" Problem

As one might expect, direct evidence rarely exists that communities of color are targeted for the siting of hazardous facilities. In fact, more recent regional studies suggest an intricate mix of social phenomena underlying the racial disparities. In the Los Angeles area, for example, the choice for siting hazardous projects usually is a community experiencing a demographic shift from one ethnic minority to another, which weakens the social ties and investment in such communities and renders them less able to mobilize a challenge. Environmental justice advocates argue that the "chicken or egg" debate obscures how historical discrimination in zoning, in addition to siting criteria that rely upon the legacy of similar practices, produces the current inequalities.

In contrast some authorities have been more candid about targeting low-income communities. For example, a 1984 report prepared by a consultant to the California Waste Management Board advised that middle- and high- socioeconomic-strata neighborhoods should not fall within the one- and five-mile radii of a proposed site. In the international arena, the World Bank's former vice president and chief economist, Laurence Summers (now president of Harvard University), sug-

Clinton Signs Executive Order Promoting Environmental Justice

"To the greatest extent practicable and permitted by law . . . each Federal agency shall make achieving environmental justice part of its mission by identifying and addressing, as appropriate, disproportionately high and adverse human health or environmental effects of its programs, policies, and activities on minority populations and low-income populations in the United States and its territories and possessions."

"Federal Actions to Address Environmental Justice in Minority Populations and Low-Income Populations," 1994. www.ejnet.org.

gested in an internal memorandum "Shouldn't the World Bank be encouraging more migration of the dirty industries to the LDCs [less-developed countries]. . . . I think the economic logic behind dumping a load of toxic waste in the lowest wage country is impeccable and we should face up to that." Because of civil rights and constitutional legal doctrine, however, exposing the interplay between race and class is problematic—there is no right to equality on the basis of class.

Challenges Confronting Environmental Justice Activists

Despite the dismal track record of constitutional and civil rights claims in the courts and in the administrative agencies, these claims often are important components of a larger political and legal strategy to obtain relief for overburdened communities. Environmental justice advocates often use state and federal environmental laws along with media campaigns and other organizing strategies to keep future hazardous facilities out of these communities. Traditional environmental law cases have been more successful, but their drawback is that the legal framework uses the arcane terminology of pollution control requirements, not the racial, political, and economic considerations.

Challenges also have been made biased on disparities that exist in decision-making processes for environmental projects. The highly technical arena of environmental regulation is a pervasively unequal playing field. Environmental justice activists often are volunteers who are short on time, resources, or specialized knowledge to challenge decisions adequately. Activists also are prone to encounter curious obstacles in the participatory process, including lack of access to important information and hostility from regulators and other stakeholders. And this lack of access likely will worsen in years to come.

The recently enacted Data Quality Act and homeland security legislation will make it increasingly difficult for impacted communities to obtain information about the neighbors who pollute their environments. The Data Quality Act places a burden on agencies to establish the reliability of information disseminated by the government, and creates cumbersome procedures for challenging such information. Homeland security legislation may restrict information about facility accidents and emergency response plans. Moreover, Attorney General John Ashcroft's expansive interpretations of the standards for exemptions under the Freedom of Information Act are expected to hamper communities' ability to obtain vitally important information about exposures and risks that affect their lives.

Disparity in Environmental Laws

Environmental disparities also exist in agency standards, program designs, enforcement, and cleanup of contaminated properties. The EPA's standard for safe fish consumption from bodies of water meeting its standards, for example, was set with the assumption that the "average" person consumes six and a half grams of fish per day—one eight-ounce serving per month. But, as the National Environmental Justice Advisory Council noted in a report to the EPA, communities of color, low-income communities, Native American tribes, and other indigenous peoples constitute populations highly exposed to contaminants in fish, plants, wildlife, and aquatic environments. The EPA subsequently revised its standard to 17.5 grams of fish per day for the general population; however, many federal and state water quality standards now in effect still are based on the old six-and-a-half-gram allowance.

Significant potential for racial disparity also occurs in the EPA's pollution control programs. A successful challenge by the local activist group Communities for a Better Environment (CBE) to a California air pollution program revealed that a plan to reduce air pollution actually resulted in worsening air quality in predominantly Latino communities near three participating refineries. Workers at the terminals additionally faced increased risks of exposure. CBE successfully challenged the program, but the tradeable rights to pollute that were the program's primary strategy continue to be touted as the most popular and efficient way to reduce pollution overall. This is evidenced in the [George W.] Bush administration's proposed legislation of the Clear Skies Act, which would adopt a pollutions credit–trading approach to regulate dangerous toxins from power plants. The EPA and state environmental agencies have yet to adequately address the potential of these market programs to cause or exacerbate toxic hot spots in vulnerable neighborhoods.

As the 1992 *National Law Journal* article reported, racial disparities also are apparent in enforcement policies for environmental laws; but other than . . . Title VI [of the Civil Rights Act] cases that remain unresolved, there is little legal activity on this issue. In the area of contaminated properties, some communities have become actively involved in government-sponsored brownfields initiatives or have aggressively campaigned for relocation of their communities, and a few toxic tort cases have been brought. Cleanup and redevelopment of abandoned contaminated land is a difficult area because federal and state authorities have significant discretion in cleanup remedies.

More Collaboration Necessary

A thoughtful collaboration between civil rights and environmental law attorneys could result in dramatic improvements in the lives of millions of families living in toxic hot spots and contaminated areas. National priorities have shifted almost exclusively to concerns over national security, however, and critical funding sources have dried up. Wholesale and unaccountable devolution of authority to local levels accompanies increasingly narrow interpretations of civil rights and constitutional laws at all levels of the judiciary. As a result, the ability of dedicated activists to address the dire conditions in environmental justice communities is severely hampered. Environmental justice offers the civil rights bar a unique opportunity once again to unite with grassroots activists and environmental lawyers, to ensure a healthy recovery for endangered communities and increased quality of life for all.

VIEWPOINT 6

> *"You have companies and residents bringing garbage from the suburbs back into their own neighborhoods."*

Minority Community Members Sometimes Cooperate with Illegal Dumping

David N. Pellow

In the following viewpoint David N. Pellow, a professor in the Department of Ethnic Studies at the University of California–San Diego, describes a case of illegal dumping in poor African American and Latino neighborhoods in Chicago. The company involved was able to exploit the corruptibility of local officials as well as residents' poverty to make their illegal acts possible. While Pellow believes that disposal companies are at fault, he also exposes the role that minority community members themselves play in aiding illegal dumping.

As you read, consider the following questions:

1. What sort of waste was the KrisJon company dumping in the Lawndale and Austin areas of Chicago, according to the author?

David N. Pellow, "The Politics of Illegal Dumping: An Environmental Justice Framework," *Qualitative Sociology*, vol. 27, fall 2004, pp. 511–25. Copyright © 2004. Reproduced with kind permission from Springer Science and Business Media and the author.

2. What reason did a local pastor give for supporting Kris-Jon, according to Pellow?

3. According to the author, how much did KrisJon pay community members who complained about the dumping?

John Christopher was a businessman. Since the late 1980s, he was in the business of "recycling" construction and demolition (C&D) waste and finding places to dump it at the lowest possible cost. The vast majority of the waste Christopher was dumping originated from highway construction projects and remodeling firms across [Chicago's] mostly white North Side and suburbs. John Christopher began dumping his waste in working-class and low-income African American and Latino communities on Chicago's West Side, particularly Lawndale and Austin. In order to ensure that he could commit these crimes without detection by the police or City Hall, he paid local aldermen cash bribes. For example, Christopher later admitted to paying bribes in the late 1980s of approximately $5,000 per month to Alderman William Henry of the 24th Ward "in return for Alderman Henry's agreement to assist [Christopher] in using and operating the site ... without interference from the City of Chicago." Christopher's deals intensified the close links that normally exist between political institutions and economic organizations in the urban political economy.

Activists Rally Against Dumping

Every community where Christopher dumped his waste was primarily African American or Latino, as was each alderman whom he bribed. The KrisJon company claimed to be recycling the C&D waste for use in future construction operations. However, KrisJon was not actually recycling the waste. Instead, the company was crushing large rocks and concrete blocks and simply piling them up—creating dumps in the neighborhoods. Local activists soon discovered that KrisJon had no permits for this operation and was therefore in violation of several city ordinances. KrisJon was engaged in illegal dumping. . . .

Local neighborhood groups in Lawndale and Austin protested against the illegal dumping operations early on. In 1990, residents held public hearings to discuss a site that was being proposed for a dump that KrisJon was to operate. A short while later, when the site was operational and producing a large volume of dust, an organization called Concerned Parents of Sumner, Frazier, and Webster Elementary School Children (schools located within blocks of the dump) sent letters to John Christopher requesting a number of environmental improvements. Christopher never responded. His dumpsites, however, were bustling and receiving waste from ninety-six different locations around metropolitan Chicago. The noise and vibrations from trucks delivering the waste was so extensive that it cracked the streets and sidewalks, damaged the foundations of nearby homes, and kept residents awake at night. Residents believed that the dust from the operation was linked to severe respiratory problems in these African American and Latino communities. . . .

Some Community Members Support the Company

The resident-activist stakeholders were fighting what appeared to be either a case of near total insensitivity by the state or an alliance between government and polluters to allow the dumps to remain in these communities of color.

But not everyone in the community was opposed to the dumping operations. Some residents joined John Christopher and participated in neighborhood "beautification projects" wherein he provided them with grass, flower, and vegetable seeds for landscaping and gardening. He also received the support of one of the strongest institutions in the African American community—the church. In 1991, a local pastor wrote a public letter applauding Christopher's efforts to "give back" to the community, and thirty residents signed a petition indicating they ". . . welcome the KrisJon Construction Company into the community and are grateful that the company is involved in a beautification project that will benefit the community and its residents." Hence even under these particularly pernicious circumstances, there were local leaders willing to publicly support even the most egregious violations of community sentiment.

While local leaders and politicians were involved in this struggle, the federal government was also well aware of the illegal dumping operations and used them as an opportunity to launch an investigation of political corruption. In 1992 the Federal Bureau of Investigation (FBI) secretly secured John Christopher's cooperation in what became known as Operation Silver Shovel. Christopher became a "mole," working undercover for the U.S. Attorney's Office and the FBI in an effort to uncover political corruption associated with the disposal of solid waste in Chicago. So, just as he had done in the 1980s, Christopher bribed African American and Latino aldermen to allow him to dump waste in their wards; the only difference was that in 1992 he was secretly videotaping the transactions. The public was not informed about this sting operation until 1996 when the media broke the case. . . .

Inequalities: Race, Class, and Political Power

While local aldermen were receiving bribes to allow dumping to continue unabated, activists were aware of the environmental inequality/racism dimensions of this struggle. One resident wrote a letter to the Cook County State's Attorney about KrisJon:

> The company operates *only in minority areas*. We also know that the company poses health hazards, damages our buildings and houses, and decreases our property values.

Echoing this sentiment, a local newspaper in the Austin community referred to the dumping practices as

> ..."environmental racism" . . . in poor areas of Chicago where toxic dumpers, solid waste treatment companies and others see large profits to be made from garbage and industrial waste [and they] attempt to gain footholds with the cooperation of greedy, corrupt, and stupid elected officials. Apparently, the first battle-ground is North Lawndale.

This writer makes two major points. The first is that the dumpsites are in poor communities; the second is that certain elected officials facilitated this process. I will address both issues because they underscore the nature of social inequality in these communities.

Poverty and Corruption

The two largest illegal dumpsites on the West Side were on Kostner and Kildare Avenues. The level of class and racial inequality evident at these sites is remarkable. For example, the neighborhood surrounding the Kostner dump (based on a one-square-mile radius from the site) had a median household income of $20,469 compared to a citywide median income of $26,301. The majority of residents in this community are people of color, with Latinos comprising 46.3 percent and African Americans 40.2 percent. Citywide, African Americans comprise 39.1 percent of the population and Latinos 19.6 percent. Thus the median household income in this neighborhood is well below both the national poverty level and the citywide median, and the percentage of people of color is higher than the citywide percentage. The figures for the Kildare site are even starker. For example, the neighborhood surrounding the Kildare dump (based on a one-square-mile radius from the site) had a median household income of $15,113. The majority of people in this community are also people of color, with African Americans comprising 89.6 per-

Aiding Illegal Dumpers for Bribes

Chicago-style corruption is at the core of this illegal dump [on the predominantly African American West Side] and several others around town. Cash for trash, it has been called. A former Chicago alderman, Bill Henry, allegedly sold out his community by taking $5,000 a month in bribes allowing the dumping to start. Mr. Henry was already facing other corruption charges in 1992 on separate issues, and he died before being indicted for the illegal dumping. At least half a dozen current or former aldermen have been named in the probe. One of them, Ambrosio Modrano, has already pleaded guilty and resigned. As many as 40 other lower-ranking city officials could be indicted for corruption connected to the illegal dumping.

Shirley Jahad, from "Chicago Dumped on by Corruption," transcription of Living on Earth, *February 16, 1996. Copyright © 1996 by World Media Foundation. Reproduced by permission.*

cent and Latinos 6.6 percent. Both of these communities are what [sociologist] William Julius Wilson has called "new poverty areas," where the majority of people live in deep poverty and most adults are either unemployed or underemployed. Thus, the majority of the residents in these neighborhoods experienced a significant degree of poverty, economic instability, and relative deprivation.

With regard to the role of corrupt elected officials, this dimension of inequality requires analysis but should also include a consideration of activities by residents who themselves accepted bribes and facilitated illegal dumping practices. The relative lack of status and political influence over citywide politics among Chicago's African American and Latino aldermen is rivaled only by the political powerlessness among their constituents. Poor and working-class residents of color in Chicago are not particularly influential in local politics and therefore offer polluters easy targets. The attendant lack of economic stability that characterizes many of these neighborhoods only reinforces their political powerlessness and the diminished status of their elected officials. This subjugated position also renders these groups particularly vulnerable to efforts by polluting firms to "divide and conquer" residents over potential economic benefits that may accompany industrial activity.

Bribes Paid to Residents

Frequently, when communities confront environmental threats, they are beset by internal fractures surrounding family conflicts, fear of job losses, and loyalties to various neighborhood institutions and firms involved. These tensions generally intensify the pain and anxiety that normally develop during conflicts over environmental contamination. Chicago's West Side was characterized by a particularly intense array of divisive wedges that rendered these working-class, polluted communities of color quite vulnerable.

Foremost among these fractures was the abuse—by aldermen—of their political positions to allow waste dumping in return for cash. But in addition to this, John Christopher had to build broader community support for his operation to ensure its survival. One strategy was to bribe *residents* who had complained about his facilities. One resident reported:

> The [truck] running up and down the street shaking the building so bad it cracked the front of the building and the roof have come loose.... My son has shortness of breath especially when the wind is coming from the west. I told John [Christopher] that his dust was a problem covering my cars every day. So he gave me $20. I told him that wouldn't pay for a whole year. Earlier he had sent $15.

For some observers fifteen or twenty dollars may seem outrageously small for a "pay-off." However, in some West Side neighborhoods in Chicago, including Lawndale and Austin, the unemployment rate among African American residents has exceeded 50 percent in the last decade. Given this context, the low price for compliance is not so surprising.

Street Gangs Help Illegal Dumpers

While some segments of the community were united against illegal dumping, the fractures remained. The divisions and betrayals within the West Side communities went much deeper than the misdeeds of corrupt politicians, immoral businessmen, and the occasional resident in need of cash. In fact, local street gangs in the Austin community were discovered regulating the fly dumping trade and were said to have charged as low as five dollars per ton of waste dumped on vacant lots. In another instance, in December of 1997, a number of Austin residents employed by a North Side remodeling company were cited for illegally dumping debris in their own community. "You have companies and residents bringing garbage from the suburbs back into their own neighborhoods," one observer commented.

Austin and Lawndale are two communities in desperate need of sustainable economic development, so politicians and residents were easy prey to bribes, temporary jobs, or a range of cash-producing activities associated with illegal waste dumping. This dynamic illustrates the depths of economic despair in many communities of color, which have become so desperate for development that garbage—or one's willingness to accept it—is viewed as one of the only marketable resources available.

Periodical Bibliography

The following articles have been selected to supplement the diverse views presented in this chapter.

Margaret Ambrose "David Suzuki: We Can Change the World in One Generation," *Habitat Australia*, October 2006.

BioCycle "The State of Garbage in America," April 2006.

Susan Buckingham, Dory Reeves, and Anna Batchelor "Wasting Women: The Environmental Justice of Including Women in Municipal Waste Management," *Local Environment*, August 2005.

Howard Frumkin "Health, Equity, and the Built Environment," *Environmental Health Perspectives*, May 2005.

Jennifer Hattam "Green Streets," *Sierra*, July/August 2006.

Rich Heffern "From Landfills to Freeways: Movement Links Ecology, Justice," *National Catholic Reporter*, June 16, 2006.

Pat Joseph and Michelle Garcia "'Race and Poverty Are Out of the Closet' (Sociologist Robert Bullard Talks About Environmental Justice)," *Sierra*, November/December 2005.

Ted May "The Sustainability Revolution," *Journal of Environmental Education*, Summer 2006.

Jim Motavalli "Trashing the Greens: Reports of Environmentalism's 'Death' May Be Exaggerated," *E: The Environmental Magazine*, May/June 2005.

Judith Petts "Enhancing Environmental Equity Through Decision-Making: Learning from Waste Management," *Local Environment*, August 2005.

Matt Watson and Harriet Bulkeley "Just Waste? Municipal Waste Management and the Politics of Environmental Justice," *Local Environment*, August 2005.

CHAPTER 2

Is Recycling Environmentally and Economically Successful?

Chapter Preface

In 1987 the garbage barge *Mobro 4000* wandered thousands of miles along the Atlantic coast and through the Gulf of Mexico seeking a place to offload its cargo. The ship, which was hauling more than three thousand tons of refuse from New York State, became a symbol of a national "garbage crisis." News reports indicated that the *Mobro's* plight was a symbol of America's shortage of landfill capacity in the face of an ever-increasing amount of trash. In response, concerned citizens prompted local governments to introduce mandatory curbside recycling as a way of reducing the amount of waste that needed to be dumped into landfills.

In the 1990s as more and more communities instituted compulsory sorting, separating, and collection of various types of garbage, a number of voices were raised in opposition to such programs. "Recycling does sometimes makes sense—for some materials in some places at some times. But the simplest and cheapest option is usually to bury garbage in an environmentally safe landfill," wrote journalist John Tierney in a 1996 essay titled "Recycling—Is Garbage." Tierney pointed to the money spent on the program—"Every time a sanitation department crew picks up a load of bottles and cans from the curb, New York City loses money"—as an indication of its inefficiency. Tierney's essay indicated the beginning of a rethinking of the benefits of recycling.

Opponents of mandatory recycling based their case on the efficiency of the free market to decide which products should be recycled. They noted that if a product could be recycled at a profit then private enterprise would step in to do the recycling. The fact that recycling efforts had to be subsidized meant that they were economically inefficient. Economic inefficiency could indicate environmental inefficiency—more energy and labor were going into the effort of collecting, sorting, and processing recycled goods than would be otherwise spent in simply burying the trash and producing new goods from "virgin" materials. "Sending around city trucks to pick up glass, paper, and plastic actually consumes more energy than it saves. And it may even pollute the air more than simply pitching this stuff," wrote the editor of *Machine Design* magazine.

Recycling advocates dispute the view that recycling requires more energy and leads to more environmental damage than landfilling. Responding to Tierney's article, Richard Denison and John Ruston of the Environmental Defense Fund wrote that "materials collected for recycling have already been refined and processed once, so manufacturing the second time around is usually much cleaner and less energy-intensive than the first. At current recycling levels, for example, the United States is saving enough energy through recycling to provide electricity for 9 million homes." In most areas, producing materials from recycled products leads to less waste, according to Denison and Ruston. They accused Tierney of picking only data that supported his argument while ignoring facts that would show recycling in a favorable light.

The "environmental accounting"—figuring out the exact costs and benefits of recycling programs—is a difficult task; it is understandable that there were and continue to be debates over the benefits of recycling. After a decade of efforts, however, there are indications that communities are learning from experience how to make their recycling programs more efficient. At the same time there are new challenges, such as the increasing amount of obsolete electronic goods ending up in municipal waste streams. In the following chapter, authors take positions on the recycling question, ranging from strictly "anti" to more nuanced analyses of how and where recycling can be successful.

"For communities in which recycling remains viable, integration, education and long-term contracting are the tools of success."

Municipalities Can Profit from Well-Designed Recycling Programs

Kivi Leroux Miller

The following viewpoint by environmental writer Kivi Leroux Miller uses interviews and data from city officials to make the case that recycling programs can benefit municipalities. The various cities profiled have differing approaches to operating waste collection and recycling services; some judge success by participation rates while others look to ensure that recycling is an economic benefit to the municipality. According to Miller, as programs evolve and officials gain more experience in running them, the benefits will continue to grow. This growth is necessary for their survival in an era of budget cuts, she maintains.

As you read, consider the following questions:

1. Why is Sunnyvale, California's, recycling program not vulnerable to budget cuts, according to Miller?

Kivi Leroux Miller, "Is Recycling Disposable?" *American City & County*, vol. 117, iss. 7, May 2002. Copyright 2002 by Prism Business Media. Reproduced by permission of the author.

2. As the author reports, how has Baltimore ensured that it won't pay more to recycle its waste than to put it in a landfill?

3. Why will the city of Albuquerque not pick up glass for recycling, according to Miller?

As local government revenues continue to shrink, city and county leaders are searching for ways to balance their budgets, often by scaling back the services they provide to their residents. Curbside recycling is one of the many services scrutinized during financial hard times, but few officials are willing to trash it. "Recycling is competing with other big policy issues, but it's a service that people have come to expect, and they would be angry if it disappeared," says David Robinson, recycling coordinator for Philadelphia.

Recycling coordinators in cities big and small, with recycling rates high and low, say the same thing: Their residents want curbside recycling, and they expect their local officials to make sure they get it. "I joke that if we eliminated the curbside program as a cost-cutting measure, a mob would descend on City Hall and hang me from the flag pole," says Mark Bowers, solid waste program manager for Sunnyvale, California. [He adds,] "It's not too far from the truth, I think."

Growth in Recycling Programs Is Slowing

Unlike other government-subsidized services, such as drinking water and wastewater treatment, recycling does not produce readily visible benefits, making it an easy target for cuts. Nevertheless, the economic and environmental benefits are real.

A study [in 2001] by the Alexandria, Virginia–based National Recycling Coalition found that the recycling industry employs 1.1 million people nationwide, generating an annual payroll of $37 billion and grossing $236 billion in annual sales. Those figures translate into a substantive revenue stream for local governments. Furthermore, the U.S. Environmental

Protection Agency has presented results of studies demonstrating that recycling saves natural resources and energy.

Because of those benefits, curbside recycling continues to grow in the United States, albeit much more slowly than it did in the boom times of the late 1980s and early 1990s. According to Jerry Powell, editor of *Resource Recycling* magazine, more than two dozen U.S. communities—including Tampa, Florida; Nashville, Tennessee; and San Diego, California—have launched new curbside programs or expanded existing ones. . . .

At the same time, seven programs with low participation in Alabama, Kentucky and Texas ceased operations. [In 2002], cities such as Baltimore and Albuquerque, New Mexico, are rearranging or cutting back their schedules to better consolidate pickups.

Integration Can Help Keep Costs in Check

To help their curbside programs get off the ground and thrive, local governments are relying on a variety of tools. For example, recycling coordinators are integrating recycling into their solid waste management systems, emphasizing public education and outreach, and establishing long-term contracts to hedge against the sometimes wild swings in recycling markets.

In 1990, Baltimore County, Maryland, integrated curbside recycling into its solid waste program by substituting one trash day with one recycling day. For each of 230,000 households, the county took away one of two weekly trash pickups and added once-a-week curbside pickup of recyclables.

The move held the county's collection costs roughly in check, says Charlie Reighart, recycling coordinator for the county. "A program like ours is insulated somewhat from the charge that recycling is an expensive proposition," he notes.

"We've stuck by the decision [to remove the second trash collection day] and never looked back," Reighart says. Today, the county's combined rate for commercial and residential recycling is approximately 40 percent.

Like Baltimore County, Madison, Wisconsin, operates its curbside recycling as part of an integrated collection system. "The city collects everything, so we can look at our whole system at once," says George Dreckman, the city's recycling coordinator.

By using the same trucks to handle trash pickup as it does to handle recycling pickup, Madison has been able to eliminate six garbage trucks and the costs of owning and operating them. According to Dreckman, communities that operate recycling separately from other waste collection activities—or those that privatize that part of their systems—do not realize those savings.

Integrated programs have other payoffs, as Bowers discovered in a detailed analysis of Sunnyvale's costs for curbside collection and other elements of the city's waste management system. In evaluating the program's performance, Bowers looked at the cost of collecting and processing materials, the revenues from the sale of materials, and the impact of diversion on tipping fees and landfill charges. The city's net diversion costs are $128 per ton for curbside recyclables; $118 per ton for yard trimmings; and $38 per ton for cardboard.

Sunnyvale's source separation programs—including curbside and yard trimmings collections and mixed-waste processing at the city's materials recovery facility—are necessary for the city to surpass California's 50 percent diversion requirements, Bowers says. As a result, curbside recycling is a fixture, protected from the cuts that plague financially tight times.

Reaching Out for Citizen Participation

Local governments that struggle to boost curbside recycling rates often attribute their difficulties to lack of staff energy, a lack of commitment to increasing participation rates over the long term and inconvenient service for residents. According to

David Robinson, recycling coordinator for Philadelphia's Streets Department, outreach also affects a program's performance.

Like Baltimore County, Philadelphia has provided curbside recycling for more than 10 years, yet its residential recycling rate has never surpassed 7 percent. Robinson attributes that, in part, to an outreach program that had historically failed to garner attention. "You need an exciting program, not just brochures and door hangers," he notes.

Robinson's office launched a new outreach campaign in March [2002] to grab residents' attention. Using television, radio and print advertising, the campaign urges residents to comply with the city's 1987 ordinance requiring participation in the recycling program.

Robinson is optimistic that the advertising—combined with improvements in collection efficiency and customer service, will bring about the behavioral changes he seeks. While the city's goal is to reach a double-digit recycling rate in three years, Robinson's goal is for the city to reach three times the current rate in that time.

Ensuring Good Prices for Recyclables

Although some communities measure recycling success by participation rates, many base their judgment on economic performance. The economics of curbside recycling are subject to unpredictable market swings, and, as a result, a recycling program that makes money one year may lose money the next. By signing long-term contracts with processors, local governments can moderate the ups and downs.

In Baltimore, city crews collect commingled recyclables from approximately 203,000 households. Those materials are processed by private firms that have recently signed long-term contracts with the city.

As of March 2002, Baltimore received $11.50 per ton of mixed paper, which includes corrugated containers and news-

Promoting Recycling to the Community

Much of what we are accustomed to throwing away can be easily put to use. A willingness to separate household wastes is all that is required. Aluminum, glass, paper, and plastics are the most useful recyclables at present. A simple system of sorting can be easily adapted to everyone's home.

Government and community leaders now realize that recycling has significant benefits. The most appealing aspect is economic. Resource recovery saves money, creates jobs, and conserves our dwindling natural resources. Garbage is a source of valuable assets that can be uncovered by all of us. Purchasing products made from recycled materials helps keep the process successful. By recycling these ordinary throwaways we can contribute to the wealth and well-being of ourselves and our world.

Balitimore, Maryland, Department of Public Works.
www.ci.baltimore.md.us/government/dpw/recycle.html.

print, from its paper processor. The city paid $11.50 per ton to another processor to handle the city's glass, metals and plastics. . . . Because the city collects more fiber tonnage than it does mixed containers, it earns a net revenue on the sale of recyclables.

According to Thompson, the city's contracts are structured to ensure that recycling is never more expensive than disposal. The agreements dictate that the city will never pay more than $34 per ton to process the recyclables, which is the amount it costs the city to dispose of a ton of garbage at the nearby waste-to-energy facility. "With these new contracts, recycling is helping us with the budget for the first time," Thompson says. "It will help us in prolonging all of the services the city offers."

Collecting Only High-Value Materials

Other local governments ensure net revenue by limiting the types of material they collect. For example, Albuquerque collects office paper, newspaper, corrugated containers, steel and aluminum cans, and HDPE and PET plastics at the curb; but residents who want to recycle glass must take their bottles to one of 15 drop-off locations around the city. "We take a conservative outlook," says Will Hoffman, training specialist for the city's Solid Waste Management Department. "We only collect materials curbside with stable, sustainable markets."

By diligently following that approach, the city has kept its recycling fee under $2.00 per month per household, Hoffman says. (That figure represents a 30 percent increase over the original fee established [in the early 1990s].) The city diverts approximately 20 percent of its waste stream, and Hoffman expects that volume to remain steady.

Although Albuquerque is comfortable with selective collection, the cost-control method is not feasible for every community. For example, cities required to meet local or state diversion levels usually need to collect materials such as glass and certain grades of paper, even if markets for those materials are poor.

Further Opportunities for Growth

As communities work to ensure growth of their curbside recycling programs, coordinators are hoping for state and federal help—especially in developing markets for the sale of recycled material and in developing consumer outreach and education programs. According to Kate Krebs, executive director for the National Recycling Coalition, the federal government's buy recycled policy will ensure a market to some degree. However, she notes that help most likely will come in the form of voluntary initiatives, such as those promoted by the National Electronic Product Stewardship Initiative (NEPSI). Organized by the Center for Clean Products and Clean Technologies at the University of Tennessee, Knoxville, NEPSI brings public and private stakeholders together to develop solutions for collecting, reusing and recycling electronic equipment.

Even as they wait for new markets to develop, local governments will continue to field demands by residents for recycling services. For communities in which recycling remains viable, integration, education and long-term contracting are the tools of success.

> *"Curbside recycling costs between 35 and 55 percent more than simply disposing of the item. It typically wastes resources."*

The Benefits of Recycling Are a Myth

Daniel Benjamin

Daniel Benjamin is a professor at Clemson University. In this viewpoint, he lays out what he calls the "eight great myths of recycling." Benjamin holds that contrary to popular wisdom, recycling does not save resources. In fact, recycling often requires more labor and fuel than producing goods from raw materials. Therefore recycling efforts waste resources that would be better spent elsewhere. By doing away with mandatory recycling, society can let the free market decide which goods make economic and environmental sense to recycle.

As you read, consider the following questions:

1. What are the three ways of dealing with the by-products of production and consumption, according to the author?

Daniel Benjamin, "The Eight Myths of Recycling," *The American Enterprise*, vol. 15, Jan.–Feb. 2004. Copyright 2004 American Enterprise Institute for Public Policy Research. Reproduced with permission of *The American Enterprise*, a national magazine of politics, business, and culture (taemag.com).

2. If landfills are constructed according to Environmental Protection Agency regulations, how many additional cancer-related deaths are expected over the next three hundred years, according to Benjamin?

3. How many garbage trucks does the author say that Los Angeles will have to add in order to implement its recycling program?

Garbage is the unavoidable by-product of production and consumption. There are three ways to deal with it, all known and used since antiquity: dumping, burning, and recycling. For thousands of years it was commonplace to dump rubbish on site—on the floor, or out the window. Scavenging domestic animals, chiefly pigs and dogs, consumed the edible parts, and poor people salvaged what they could. The rest was covered and built upon.

A Manufactured Crisis

Eventually, humans began to use more elaborate methods of dealing with their rubbish. The first modern incinerator (called a "destructor") went into operation in Nottingham, England in 1874. After World War II, landfills became the accepted means of dealing with trash. The modern era of the recycling craze can be traced to 1987, when the garbage barge *Mobro 4000* had to spend two months touring the Atlantic and the Gulf of Mexico before it found a home for its load. The Environmental Defense Fund, the National Solid Waste Management Association (whose members were anxious to line up new customers for their expanding landfill capacity), the press, and finally the Enviromnental Protection Agency [EPA], spun the story of a garbage crisis out of control. By 1995, the majority of Americans thought trash was our number one environmental problem—with 77 percent reporting that increased recycling of household rubbish was the solution. Yet these claims and fears were based on errors and misinformation, which I have compiled into the Eight Great Myths of Recycling.

Myth 1: Our Garbage Will Bury Us

Fact: Even though the United States is larger, more affluent, and producing more garbage, it now has more landfill capacity that ever before. The erroneous opposite impression comes from old studies that counted the number of landfills (which has declined) rather than landfill capacity (which has grown). There are a few places, like New Jersey, where capacity has shrunk. But the uneven distribution of landfill space is no more important than the uneven distribution of automobile manufacturing. Perhaps the most important fact is this: If we permitted our rubbish to grow to the height of New York City's famous Fresh Kills landfill (225 feet), a site only about 10 miles on a side could hold all of America's garbage for the next century.

Myth 2: Our Garbage Will Poison Us

Fact: Almost anything can pose a theoretical threat, but evidence of actual harm from landfills is almost non-existent, as the Environmental Protection Agency itself acknowledges. The EPA has concluded that landfills constructed according to agency regulations can be expected to cause a total of 5.7 cancer-related deaths over the next 300 years. It isn't household waste, but improperly or illegally dumped industrial wastes that can be harmful. Household recycling programs have no effect on those wastes, a fact ignored by messianic proponents of recycling.

Myth 3: Our Packaging Is Immoral

Fact: Many people argue that the best way to "save landfill space" is to reduce the amount of packaging Americans use, via mandatory controls. But packaging can actually reduce total garbage produced and total resources used. The average American family generates fully one third less trash than does the average Mexican household. The reason is that our intensive use of packaging yields less spoilage and breakage, thereby saving resources, and producing, on balance, less total rubbish. Careful packaging also reduces food poisoning and other health problems.

Over the past 25 years, market incentives have already reduced the weights of individual packages by 30 to 70 percent. An average aluminum can weighed nearly 21 grams in 1972; in 2002, that same can weighs in at under 14 grams. A plastic grocery sack was 2.3 mils [millimeters] thick in 1976; by 2001, it was a mere 0.7 mils.

By contrast, the environmentally sensitive *New York Times* has been growing. A year's worth of the newspaper now weighs 520 pounds and occupies more than 40 cubic feet in a landfill. This is equivalent in weight to 17,180 aluminum cans— nearly a century's worth of beer and soft drink consumption by one person. Clearly, people anxious to heal Mother Earth must forego the *Times*!

Myth 4: We Must Achieve "Trash Independence"

Fact: Garbage has become an interstate business, with 47 states exporting the stuff and 45 importing it. Environmentalists contend that each state should dispose within its borders all the trash produced within its borders. But why? Transporting garbage across an arbitrary legal boundary has no effect on the environmental impact of the disposal of that material. Moving a ton of trash is no more hazardous than moving a ton of any other commodity.

Myth 5: We're Squandering Irreplaceable Resources

Fact: Thanks to numerous innovations, we now produce about twice as much output per unit of energy as we did 50 years ago, and five times as much as we did 200 years ago. Automobiles use only half as much metal as in 1970, and one optical

Skepticism About Recycling

Mandatory recycling programs aren't good for posterity. They offer mainly short-term benefits to a few groups— politicians, public relations consultants, environmental organizations, waste-handling corporations—while diverting money from genuine social and environmental problems. Recycling may be the most wasteful activity in modern America: a waste of time and money, a waste of human and natural resources.

John Tierney, New York Times Magazine, *June 30, 1996.*

fiber carries the same number of calls as 625 copper wires did 20 years ago. Bridges are built with less steel, because steel is stronger and engineering is improved. Automobile and truck engines consume less fuel per unit of work performed, and produce fewer emissions.

To address the issue of paper, the most-promoted form of recycling: The amount of new growth that occurs each year in forests is more than 20 times the number of trees consumed by the world each year for wood and paper. Where loss of forest land is taking place, as in tropical rain forests, it can be traced directly to a lack of private property rights. Governments have used forests, especially the valuable tropical ones, as an easy way to raise quick cash. Wherever private property rights to forests are well-defined and enforced, forests are either stable or growing. More recycling of paper or cardboard would not eliminate tropical forest losses.

Myth 6: Recycling Always Protects the Environment

Fact: Recycling is a manufacturing process, and therefore it too has environmental impact. The U.S. Office of Technology

Assessment says that it is "not clear whether secondary manufacturing [i.e., recycling] produces less pollution per ton of material processed than primary manufacturing." Recycling merely changes the nature of pollution—sometimes decreasing it, and sometimes increasing it.

This effect is particularly apparent in the case of curbside recycling, which is mandated or strongly encouraged by governments in many communities around the country. Curbside recycling requires that more trucks be used to collect the same amount of waste materials. Instead of one truck picking up 40 pounds of garbage, one will pick up four pounds of recyclables and a second will collect 36 pounds of rubbish.

Los Angeles has estimated that due to curbside recycling, its fleet of trucks is twice as large as it otherwise would be—800 vs. 400 trucks. This means more iron ore and coal mining, more steel and rubber manufacturing, more petroleum extracted and refined for fuel—and of course all that extra air pollution in the Los Angeles basin as the 400 added trucks cruise the streets.

Myth 7: Recycling Saves Resources

Fact: Using less of one resource usually means using more of another. Curbside recycling is substantially more costly and uses far more resources than a program in which disposal is combined with a voluntary drop-off/buy-back option. The reason: Curbside recycling of household rubbish uses huge amounts of capital and labor per pound of material recycled. Overall, curbside recycling costs between 35 and 55 percent more than simply disposing of the item. It typically wastes resources.

In the ordinary course of daily living, we already reuse most higher value items. The only things that intentionally end up in the trash are both low in value and costly to reuse or recycle. Yet these are the items that municipal recycling programs are targeting—the very things that consumers have already decided are too worthless or costly to deal with further. All of the profitable, socially productive opportunities for recycling were long ago co-opted by the private sector, because they pay back. The bulk of all curbside recycling programs simply waste resources.

Myth 8: Without Forced Mandates, There Wouldn't Be Any Recycling

Fact: Long before state or local governments had even contemplated the word recycling, the makers of steel, aluminum, and thousands of other products were recycling manufacturing scraps. Some operated post-consumer drop-off centers.

As for the claim that the private sector promotes premature or excessive disposal, this ignores an enormous body of evidence to the contrary. Firms only survive in the marketplace if they take into account all costs. Fifty years ago, when labor was cheap compared to materials, goods were built to be repaired, so that the expensive materials could be used for a longer period of time. As the price of labor has risen and the cost of materials has fallen, manufacturers have responded by building items to be used until they break, and then discarded. There is no bias against recycling; there is merely a market-driven effort to conserve the most valuable resources.

Conclusion: Mandatory Recycling Squanders Resources

Informed, voluntary recycling conserves resources and raises our wealth, enabling us to achieve valued ends that would otherwise be impossible. Mandatory programs, however, in which people are directly or indirectly compelled to do what they know is not sensible, routinely make society worse off. Such programs force people to squander valuable resources in a quixotic quest to save what they would sensibly discard.

Except in a few rare cases, the free market is eminently capable of providing both disposal and recycling in an amount and mix that creates the greatest wealth for society. This makes possible the widest and most satisfying range of human endeavors. Simply put, market prices are sufficient to induce the trash-man to come, and to make his burden bearable, and neither he nor we can hope for any better than that.

"Recycling . . . is a geographically narrow activity, and its costs and benefits vary accordingly."

Flexibility Is the Key to Efficient Waste Disposal Practices

Douglas Clement

In the following viewpoint Douglas Clement, senior writer for the Federal Reserve Bank of Minneapolis's magazine Fedgazette, *adopts a middle-of-the-road approach to the question of recycling. Clement maintains that conditions vary widely in the United States and that different materials have different value as recyclables. Therefore, a wide variety of approaches to waste reduction is called for, he argues; landfilling might be more efficient in rural areas whereas intensive recycling might benefit urban locales. One widely applicable idea that he champions, however, is to charge households per-volume (or per-bag) fees to haul their garbage, thus giving consumers incentives to reduce and recycle.*

Douglas Clement, "Recycling—Righteous or Rubbish?" *Fedgazette*, March 2005, pp. 6–10. Reproduced by permission of Public Affairs, Federal Reserve Bank of Minneapolis.

As you read, consider the following questions:

1. According to the author, in what year did the volume of waste recycled in Wisconsin reach a peak?
2. According to Clement, what is the "critical" factor in a cost-benefit analysis of recycling programs?
3. Why might the price of materials made from "virgin" raw materials be artificially low, in the author's opinion?
4. Does charging for waste haulage through pay-as-you-throw programs reduce amounts of household waste collected, according to Clement?

Recycling has become an American norm. It's an expectation, a civic duty, almost an amenity. No one wants to take out the trash, but if we lose the opportunity or responsibility to sort through our discarded cans and newspapers, we feel deprived. . . .

But regardless of its popularity, recycling may not always be justifiable on economic grounds, and many communities are finding that it's difficult to continue their recycling programs, especially in tough economic times, when the costs of doing so seem to eclipse the benefits. A 2002 Minnesota legislative audit of state recycling efforts put it bluntly: "Progress may be limited . . . because it is often cheaper to throw away rather than recycle some materials."

Reaching a Plateau

Moreover, recycling rates seem to have leveled off or even dipped. Over 60 percent of all beverage containers were recycled in 1994, but that figure had dropped to just 47 percent in 2002. Even aluminum cans, the most lucrative recyclable in the waste stream, are missing the recycling bin, with rates down from 65 percent in 1992 to 44 percent in 2003, according to the Container Recycling Institute. *BioCycle* surveys show that overall U.S. recycling rates—the percentage of total municipal solid waste that is recycled—have been flat since

1998, with recycling rates hovering in the low 30 percent range, according to its December 2001 report.

Similar trends appear in the district. "Recycling rates have plateaued in recent years," said the January 2004 report of the Minnesota Office of Environmental Assistance. And while recyclable volumes have increased an average of 4 percent a year over the past decade, there have been "slight declines in the last two years," said the report. In Wisconsin, a December 2004 Department of Natural Resources press release trumpeted a 2 percent rise from 2002 to 2003 in pounds recycled per capita, but the figure was still 13 percent below the 1998 peak of 302 pounds per person.

Critics say such drops are a vindication of their claim that recycling is garbage—a wasteful hobby for impractical, feel-good environmentalists. But supporters claim the slowdown is just a hiccup in the continued trend toward better resource utilization. And they argue that if recycling seems more expensive than the alternatives, it's only because landfilling is artificially cheap and the use of virgin materials continues to be subsidized.

A Middle Road

The economics suggest a middle road. Careful cost-benefit analysis shows that recycling often isn't cost-effective: Many programs try too hard, in a sense, by recycling products that cost more to reprocess than is warranted by the associated environmental and economic benefits—essentially going too far in the cause of environmental protection.

But economists also suggest that some level of recycling is entirely sensible from an economic standpoint. Moreover, they say, it's a relatively young business whose systems and technologies are still developing. Costs are dropping and markets are firming up, making it an increasingly viable alternative to landfilling. Nonetheless, because the costs and benefits of recycling and other methods of waste disposal vary greatly from one area to another, blanket policies that mandate certain national or even state-level quotas make little sense. . . .

The costs of recycling programs depend on a number of factors. Collection costs are key, and there are economies of scale that make recycling a difficult proposition in sparsely populated areas like Lake County [Minnesota]. Also, it's usually cheaper to collect trash than recyclables because it can be crushed more compactly than materials that have to be processed before sale. Glass bottles, for example, need to be sorted by color; crushing them makes that next to impossible. "The collection of recyclables fills trucks more quickly and requires more trips," noted a 2001 Wisconsin [legislative] analysis.

But revenue generated by selling recyclables is often critical to the cost-benefit outcome. "That market price is really what makes or breaks recycling," said [Mike] Boerger, from Watertown [South Dakota], and markets for recyclables are notoriously volatile. "The revenue we receive averages around $9,600 or $9,700 a year [after paying the recycling contractor], but the range has swung from $3,700 to $13,700. And that's just from 1998 to 2003." The Wisconsin study reported that prices for recycled mixed paper varied in August 2000 from a high of $50 a ton to a low of -$10 (meaning sellers had to pay buyers to take the paper). Aluminum prices jumped between $532 and $1,160 a ton.

If the costs of recycling were low, revenue volatility wouldn't be a problem. But local governments are increasingly under fire for continuation of recycling programs that have dependably high costs and revenues that are unpredictable.

The Issue of Jobs

In the face of such pressure, recycling proponents muster several defenses. Recycling protects the environment, they argue, by saving resources—energy and raw materials—that are scarce and underpriced; it creates jobs and builds the economy; and it reduces the problem of rapidly diminishing landfill capacity, a problem exacerbated because people don't pay the full costs of landfill disposal. To an economist, some of these arguments have real merit while others hold no water.

The idea that recycling is good because it "creates jobs" is often voiced. "Recycling accounts for 8,700 manufacturing jobs in Minnesota, with paychecks totaling something over $1 billion," said a November 2004 Minneapolis *StarTribune* editorial. A July 2004 study from the Montana Department of Environmental Quality reported employment of 300 full-time workers with above-average wages and gross revenues for the industry of $90 million.

But to an economist, this is a red herring. After all, jobs would also be created if governments allowed citizens to throw garbage in the street and then hired people to pick it up. If tax revenues weren't paying someone to sort cans and bottles in a recycling processing facility, that money could be used in some other way, perhaps paying teachers to educate children, or never collected as a tax in the first place—both of which might have a greater social benefit.

"The fact that lots of people are needed to carry out recycling programs is basically evidence that recycling is expensive, requiring lots of labor (as well as capital) that could have been used to fulfill other goals of public policy," wrote Richard Porter, a University of Michigan economist, in his recent book *The Economics of Waste.*

Subsidies to Raw Materials Producers

Then there's the argument that recycling should be encouraged because raw materials are nonrenewable resources whose use is unwisely subsidized by misguided government policies. Moreover, goes this argument, creating new products out of recycled materials is less polluting than creating them from virgin materials.

Dubious Data

Unfortunately—and perhaps appropriately—data about garbage are quite a mess. The two most authoritative sources on the subject report very different numbers for total garbage generated in the United States.

The EPA, which estimates its figure indirectly by calculating how much waste people generate given a certain level of economic activity and then extrapolating on the basis of economic and population data, said we generated 229 million tons of MSW [municipal solid waste] in 2001.

Another respected source of waste data, "The State of Garbage" report by *BioCycle* magazine and Columbia University, goes after the figure more directly: It asks state officials how much waste their residents generate. The most recent report (January 2004) reduces the flaws often found in MSW data by accounting for trash traded among states and by excluding construction and demolition debris and industrial waste. Even trimmed of this fat, their MSW figure for 2002 is far higher than the EPA 2001 total: 369 million tons.

Fedgazette, *March 2005. www.minneapolisfed.org.*

Resource economists generally agree that resource extraction has long been subsidized by government policies—tax depletion allowances, and the like. One academic study estimated, for example, that aluminum production in the United States receives a subsidy of roughly 5 percent to 12 percent because much of it is produced with low-cost, government-run hydroelectric power. But subsidizing recycling to redress this would be a second-best policy, trying to right one wrong by committing another. The better policy would be simply to eliminate unjustified subsidies.

As for growing shortages of raw materials, economists would argue that accurate pricing and unfettered markets are the only efficient way to address whatever scarcity might really exist. If raw wood pulp prices double, you can be sure that paper makers will create a market for recycled paper, develop a technique that uses less pulp or plant more trees.

And if production processes with virgin materials cause pollution, that externality should be taxed directly. Supporting production out of recycled goods is an indirect means of addressing the pollution. Moreover, recycling itself is not perfectly "green." Washing out peanut butter jars uses lots of water. Trucks that collect recyclables pollute the atmosphere. And while few materials are as recyclable as aluminum, that too causes pollution. In November 2004, for instance, the Environmental Protection Agency proposed a $247,578 penalty against a Minnesota aluminum recycling plant for emitting "excessive amounts of dioxins and furans and hydrochloric acid." Nasty stuff, that recycling.

Pay As You Throw

But another point has much greater credibility. The time-consuming process of recycling faces an unfair fight if the alternative—tossing something in a trash can—is virtually free. And in the United States, that's often been the case. In most American municipalities, the standard practice has been for households to pay for trash collection through taxes assessed on the market value of their homes and/or with a monthly garbage bill unrelated to the amount of trash put out for disposal. As a result, an individual homeowner (or business, in many cases) faces essentially no marginal cost for throwing away another bag or three of trash each week. That's an obvious disincentive to recycle. Why go to the effort of sorting recyclables when you can just toss them in the garbage can for free?

In the last 20 years or so, however, many municipalities—over 4,000 according to the EPA—have instituted volume-based pricing for waste disposal, otherwise known as pay-as-you-throw [PAYT] programs. The idea of PAYT is that people should be charged for garbage disposal the same way they're charged for gasoline or bananas: by the amount they consume. Some communities sell colored tags that residents attach to standard-size garbage bags, usually about 33 gallons. The purchase price of the tag is the unit price they pay for having that garbage hauled to the landfill. Others send residents garbage bills that vary by the size of their trash container. The bigger the can, the higher the bill.

The city of Bozeman [Montana] implemented its PAYT program in 1991, initially using a tag-and-bag system, then switching to garbage cans of different sizes. And it would be hard to find anyone more enthusiastic about the plan than Mark Kottwitz, the city's solid waste superintendent.

"Our philosophy behind the pay-as-you-throw or volume-based system would be whoever generates the most pays the most," he said. "The whole incentive behind going to a smaller container is our recycling program. The more you recycle, the smaller [the] container you can go to." A 35-gallon trash can emptied once a week costs a resident $8.77; a 100-gallon can costs $14.35. "The program's really working well," said Kottwitz. "The community is just doing a fabulous job."

Illegal Dumping

To an economist, pay-as-you-throw makes perfect sense. In a competitive economy, people should be charged for the goods and services they consume, and the price should equal the marginal cost. But there can be unintended consequences.

Terry Barnes, former chairman of the Upper Peninsula [of Michigan] Recycling Coalition, likes PAYT programs. "If a city doesn't have a bag fee, there's no incentive to recycle," he observed. And when nearby Florence, Wisconsin, started its recycling plan together with a PAYT trash program in the early 1990, "garbage volume dropped by 50 percent." But diversion to recycling "only represented about 14 percent of the loss," he recalled. "The other part of it went to cities across the river like Iron Mountain [Michigan]" where Florence citizens dropped off their trash bags next to Iron Mountain homes. . . .

Think Locally

Still, recycling isn't an all or nothing proposition. It may not make much sense to push recycling in sparsely populated eastern Montana or northern North Dakota where landfills are near and markets distant. Sioux Falls and Minneapolis, on the other hand, might benefit from accelerated recycling efforts. And programs may be wiser to focus on increased collection of some materials—aluminum and paper, for instance—but drop others, like some types of plastic (and maybe even plastic bags) that are expensive to collect and process, and have little market value.

But efforts to impose uniform recycling targets over broad geographic areas or to mandate collection of an extensive range of materials are doomed to inefficiency. As noted in the 2002 Minnesota legislative audit of that state's recycling efforts, "each county has a unique political, social and geographic environment that might require a unique solution to waste management." Recycling, when it comes down to it, is a geographically narrow activity, and its costs and benefits vary accordingly. Like politics, all garbage is local.

"The value of the materials recovered from curbside recycling is far less than the extra costs of collection, transportation, sorting, and processing."

Curbside Recycling Wastes Environmental and Economic Resources

Asa Janney

In this viewpoint, economist Asa Janney disputes the conventional wisdom that curbside recycling makes environmental and economic sense. According to Janney, mandatory recycling wastes the time of households who have to sort their trash and prepare separate containers for haulage. In addition, the extra energy required in gathering and processing recyclables, as well as the pollution produced by recycling, means that these types of recycling programs actually hurt the environment.

As you read, consider the following questions:

1. According to Janney, what is the "economic method" of evaluating the claims that recycling saves resources?

2. How much more per ton does New York City pay to recycle rather than bury its solid waste, as reported by the author?

Asa Janney, "Mandatory Recycling," *The Quaker Economist*, vol. 3, no. 91, December 15, 2003. Reproduced by permission.

3. What is produced when newspapers are recycled, according to Janney?

Environmentalists and the governments they influenced introduced widespread mandatory recycling. One of their claims was that recycling saves resources. I use the economic method to evaluate this claim—add up the values of the resources saved and subtract the values of the extra resources consumed and see whether the net value is positive. You have to be careful when you sift through the studies that purport to have used this accepted method; some add up the benefits and leave out costs. Particularly in the analysis of curbside recycling many studies leave out important elements, such as state and local subsidies and recycling's share of overhead. Sometimes only part of the process is examined. For example, aluminum scrap delivered to a factory that makes aluminum cans is quite valuable, but this leaves out the costs of getting the scrap to the factory. Finally, some analyses engage in double-counting. Recycling often uses less energy and raw materials. However, these features are reflected in the price of recyclable materials. Pointing out these features as extra advantages double-counts them.

Curbside Recycling Costs More than Dumping

All the careful studies I found determined that the value of the materials recovered from curbside recycling is far less than the extra costs of collection, transportation, sorting, and processing. Thus, selling the aluminum, paper, plastic, glass, etc., that are recoverable from household trash does not pay for recycling them. One factor that determines this result is that the avoided costs of trash disposal are low; even though households that practice curbside recycling send less trash to the landfill, it does not save much. Barbara Stevens found that the average avoided cost of landfill tipping fees was $7.00 per household per year. On the other side of the ledger, the extra cost of picking up household recyclables is high.

Consider the case of New York City, which loses money on its curbside recycling program. It has to pay extra administrators who run a continual public relations campaign to explain what to do with dozens of different products. You can recycle milk jugs but not milk cartons and index cards but not construction paper. The city has enforcement agents inspect garbage and issue tickets. But most of all, recycling requires extra collection crews and trucks. Collecting a ton of recyclable items is three times more expensive than collecting a ton of garbage because the crews pick up less material at each stop. For every ton of glass, plastic and metal that a truck delivers to a private recycler, the city currently spends $200 more than it would spend to bury the material in a landfill. This $200/ton includes paying $40/ton to private recyclers to take the materials because their processing costs exceed the eventual sales price of the recycled materials. More generally, the not-for-profit group Keep America Beautiful estimates that curbside recycling adds fifteen percent to the cost of waste disposal.

Rather than conducting this expensive experiment with curbside recycling, could the result that it is uneconomic have been predicted? I think so. People and households are economic agents. They reuse and recycle items within the home until their value is low and their cost to recycle is high. Then the item is discarded. So mandatory curbside recycling tries and fails to find value in items that households have declared worthless.

But suppose there were economies of scale so that someone who collected large amounts of a certain type of recyclable could squeeze some value from it? In that case, the private sector outside of households would have jumped on it. Indeed, it did. Before mandatory recycling was introduced, about ten percent of household trash was recycled. Much more, about 60 percent, of industrial waste was recycled because of the higher concentrations that lower the costs of recycling. This brings to mind another recycling myth: If recycling were not mandated, it would not happen. But indeed, it happens where it is economical. American industry voluntarily recycles 60 million tons of ferrous [iron-containing] metals, seven million tons of nonferrous metals, and 30 million tons of paper, glass, and plastic per year. These amounts tower above the totals from mandatory recycling at all levels of government.

Costs of Time to Sort Trash

Let's consider one more issue on curbside recycling. None of these estimates of the cost of recycling is complete because they do not account for the cost of our time to sort our trash and neatly arrange it in color-coded bins at the curb. Even the best private studies I quoted above only mention this cost without putting a value on it; the government studies either do not mention it or imply that we should donate our time as good citizens. John Tierney, author of "Recycling Is Garbage," measured the time it takes to comply with New York's mandatory household recycling rules and put a value of $12/hour on it—"a typical janitorial wage." At that rate, each ton collected would cost an additional $792. If you add in the rental cost of the space in a New York City home needed to store the sorted materials, the total cost of collection rises to $3000/ton. He concludes, "Recycling may be the most wasteful activity in modern America: a waste of time and money, a waste of human and natural resources."

The claim that recycling always protects the environment is another myth. Again, you have to look at the big picture. Recycling is not just stacking your newspapers at the curb; it also involves a manufacturing process which has environmental consequences. The EPA [Environmental Protection Agency] says that twelve toxic substances are found in both virgin and recycled paper processing. Eleven of these are present at higher levels in the recycling process. Steel and aluminum processing

Recycling Paper Kills Trees

If we stopped using paper, there would be fewer trees planted. In the paper industry, 87% of the trees used are planted to produce paper. For every 13 trees "saved" by recycling, 87 will never get planted. It is because of the demand for paper that the number of trees has been increasing in this country for the last fifty years. The lesson is this: if your goal is to maximize the number of trees, don't recycle.

Roy E. Cordato, Free Market, *December 1995.*

have similar mixed results. Many studies have repeatedly found that recycling can either increase or decrease pollution; it is not uniformly good for the environment.

Add to this the pollution by the extra trucks required for curbside recycling. Los Angeles found that it had to double its trash truck fleet from 400 to 800 to pick up household recycling. Consider not only the fumes those extra 400 trucks release but also the pollutants generated by processing the steel, plastics, and other materials required to build the trucks.

Environmental Effects of Recycling

Consider the environmental effects of recycling some particular products. Paper is an important one. Recycling newsprint creates more water pollution than making new paper—5000 gallons more waste water per ton. When old newsprint is recycled, every hundred tons of de-inked fiber also produces 40 tons of toxic sludge that requires special disposal. Jerry Taylor of the Cato Institute said it well: "Paper is an agricultural product, made from trees grown specifically for paper production. Acting to conserve trees by recycling paper is like acting to conserve cornstalks by cutting back on corn consumption."

Moreover, Canada makes most of our virgin newsprint using clean hydroelectric power while recycling newsprint in this country increases the consumption of fossil fuels.

How about disposable diapers, which the *New York Times* once called the "symbol of the nation's garbage crisis?" Systematic studies have found that disposable diapers make up about one percent of landfill contents, not the estimates of up to 25 percent that have been erroneously reported. Reusable diapers are not better for the environment. Using reusable diapers in favor of disposables consumes more than three times as much energy and produces ten times as much water pollution.

Reusing glass bottles consumes more energy than initial manufacture, because they consume heat for sterilization. Used bottles can be crushed and mixed with other materials to make aggregate, and glass is environmentally neutral in a landfill. . . .

The Fallacy of "Garbage Independence"

Amid all the other mandated recycling goals arose a really strange one, "garbage independence." We started to hear that a community, regardless of its size, should dispose of all its waste locally. Why should we saddle ourselves with a moral obligation to dispose of our garbage near home? Most of the goods we consume were shipped to us from factories and farms at a distance. What could be wrong with sending it out to be buried in places with open land? James DeLong, commented, "With that kind of logic, you'd have to conclude that New York City has a food crisis because it can't grow all the vegetables its people need within the city limits, so it should turn Central Park into a farm and ration New Yorkers' consumption of vegetables to what they can grow there."

The image of high-income Americans picking trash is odd. Throughout history we have had trash pickers whose opportunity costs were so low that they could afford to browse through other people's trash. When books were made from rags, rag picking was the profession of the lowest class. So, the private sector was engaged in voluntary recycling then, and we did not force those with valuable labor skills to engage in it.

"Ann Arborites would overwhelmingly reject a referendum proposal to simultaneously terminate recycling and reduce property taxes by \$4 per person a year."

A Recycling Program May Improve Social Welfare

Richard C. Porter

In the following viewpoint, Richard C. Porter, a professor of economics at the University of Michigan, conducts a cost-benefit analysis of Ann Arbor, Michigan's, recycling program. He concludes that the program actually ends up costing the city around \$401,000, or approximately \$4 per citizen; however, he believes that citizens generally accept that cost because they feel that recycling is necessary for a sustainable society.

As you read, consider the following questions:

1. What are two economic benefits of Ann Arbor's recycling program, according to Porter?
2. What does the author say is the largest single cost of Ann Arbor's recycling program?

3. What is Porter's evidence that citizens overall enjoy participating in the recycling program?

Ann Arbor [Michigan's] recycling history is typical. Starting from a program in the 1970s where enthusiastic recyclers could bring a few well-sorted recyclables to a volunteer-operated drop-off station, it has grown into a mandatory program with weekly curbside pickup and a multimillion-dollar MRF [materials recovery facility] collecting not only glass, metal, plastic, and paper but even textiles, oil filters, waste oil, batteries, ceramics, and aerosol cans. In a U.S. EPA [Environmental Protection Agency] review of 17 "record-setting" waste reduction communities, Ann Arbor was third in its recycling rate, 30%.

This [viewpoint] attempts to assign rough dollar values to the benefits and costs of the Ann Arbor curbside residential recycling effort, as of 1997. The analysis is restricted to residential recycling, omitting not only industrial and commercial efforts but also composting, which in total tonnage is even larger than recycling. The basic benefits and costs of recycling are of six kinds:

Economic Benefits of Recycling Program

Revenues from recovering recyclable materials Ann Arbor sold its nearly 14,000 tons of recyclables at an average price of $37.65 a ton, which yielded a total revenue to the city and its MRF of $517,000. (This is the revenue from all the MRF's sales of recyclables. By a complex formula, the city receives some of the revenue from the MRF's sales of collected recyclables; in 1997, the city's share was $45,000: $517,000 minus $472,000.) Assuming that the social value of the recyclables is what was paid for them, this means a social benefit of $517,000.

Costs avoided in landfill disposal Whatever tonnage of solid waste is recycled is not landfilled, and the resources needed to bury that waste are saved. Ann Arbor no longer operates its own landfill but pays $28 a ton to dispose of it at a nearby private landfill. Because Ann Arbor collects 11,586 tons a year—about two-thirds of a pound per person per day—its residential recycling program directly saves the city $324,000 a year. But the very existence of this program permits the city to add another 1,281 tons of recyclables from other [not from households] city sources, which means the program permits landfill costs to be avoided to the extent of $360,000 a year. (The drop-off center collects more than 2,000 tons, but nearly half of that is from noncity sources, so the city would not have been responsible for its landfill cost.)

Solid waste collection costs avoided Solid waste tonnage diverted to recycling also does not have to be collected as solid waste. Ann Arbor spent $626,000 in 1997 collecting 16,107 tons of residential solid waste, the average cost of such collection is about $40 a ton. If social and private cost were the same, and if the average and marginal cost [the cost for each additional ton] were the same, this would mean that recycling reduces the total social cost of collecting MSW [municipal solid waste] by $450,000. Social and private cost probably are about the same, but the marginal cost of collecting an extra ton of MSW is almost certainly less than the average cost. It is not easy to know how much less. Assuming marginal cost is half of average cost yields a social benefit on this account of $225,000.

Costs of Recycling Program

Recyclables collection costs incurred The city contracts out the curbside residential collection of recyclable materials at a cost to the city of $1,014,000. Assuming that nonprofit contractor Recycle Ann Arbor breaks even on this activity, it costs about $87 a ton, for a social cost of $1,014,000.

Operating costs of the MRF The MRF is also contracted out by the city, and the contractor gives out little information on its finances. But, again assuming that its total private revenues

Cost-Benefit Analysis of Recycling in Ann Arbor, Michigan, 1997	
Category	Cost
Revenue from Recyclable Materials	+$517,000
Landfill Costs Avoided	+$360,000
MSW Collection Costs Avoided	+$225,000
Recyclables Collection Costs Incurred	−$1,014,000
Operating Costs of the MRF	−$739,000
Transfer Costs Avoided	+$250,000
Net Total	−$401,000

TAKEN FROM: Richard C. Porter, *The Economics of Waste*. Washington, DC: Resources for the Future, 2002, p. 145.

equal its total social costs, we can add together what the MRF gets from the recyclable revenues ($472,000) and what the city pays the MRF for its activities ($267,000) to get an estimate of its social cost of operation: $739,000, or just above $63 per ton of recyclables processed. Technically, because the MRF is already built, such a sunk cost should not count in a benefit-cost analysis. What the city pays for the MRF undoubtedly does include both interest and depreciation, so this analysis is counting these sunk costs as if they had not yet been undertaken.

Transportation Costs

Transfer cost avoided In Ann Arbor, there is another benefit to recycling. . . . Without recycling, all of Ann Arbor's trash would have to be moved by the collection trucks to the private landfill 25 miles away. With recycling, all the trash goes only as far as the nearby MRF. What is not recycled is then transferred and further compacted for its lower-cost trip to the landfill. As a result, recycling saves about 250,000 miles a year of travel for its trash collection trucks. At $1 a mile, this means a benefit of $250,000. Should Ann Arbor have considered a transfer station without recycling? We cannot answer this question, because there is no easy way to disentangle the transfer costs from the recycling costs at the MRF. (The costs of the compacting at the transfer station and of the semi [trailer truck] that moves the waste to the landfill are included in the costs of the MRF.)

Ann Arbor's recycling program currently [in 2002] has a net social value of about minus $400,000 per year. Does this mean that recycling in Ann Arbor should be abandoned? No— there are four important other considerations:

- Some of the numbers are only estimates, and two in particular are iffy. First, prices of recyclable materials in 1997 were very low by historical standards. The revenues in the next year, 1998, for example, were more than $1 million. What is important for this benefit-cost analysis is the average price of recyclable materials over the life of the MRF. Second, the estimate of the savings in solid waste collection due to a separate recycling collection—$225,000—*assumed* that the marginal cost of trash collection is half of average cost. In fact, we can only be sure that something between $0 and $450,000 was saved in municipal trash collection. And whatever was saved in 1997, even if we knew it exactly, would only provide a lower-bound estimate of long-run annual savings, as routes, labor, and trucks were adjusted to the new situation and "learned" how to handle the process more cheaply.

- We have not counted either the costs or benefits of recycling to the households that put their home effort into the program. For some, recycling is a pain that they suffer because it is mandatory; for others, it is a joy that offers them a chance to contribute to a more sustainable society. The dollar values of these pains and joys are hard numbers to estimate; but if they net out to an average benefit of more than $4 per Ann Arborite per year, then recycling does pass its social benefit-cost test. Certainly, Ann Arborites would overwhelmingly reject a referendum proposal to simultaneously terminate recycling and reduce property taxes by $4 per person a year.

- Even if recycling now fails its benefit-cost test, there may be learning by doing, and the costs of recycling may be coming down. It is the cost over the life of the MRF that matters, not just the 1997 cost.

- Recycling may fail its benefit-cost test because Ann Arbor is recycling too much. Minor items—such as ceramics, textiles, #3 plastic, batteries, oil filters, and waste oil—raise the sorting costs at the MRF by much more than they raise revenues: But notice the examples here; some—such as batteries, oil filters, and waste oil—are collected not so much to recycle them as to keep them from being carelessly discarded and contaminating groundwater. We should be counting this expected groundwater pollution avoided as a benefit of the recycling program. Still, the inclusion of many minor items in Ann Arbor may tilt its overall benefit-cost result from positive to negative.

One final point, following from the fourth consideration just above. We get a clue about the marginal cost of minor items by looking at a private trash collection and recycling company operating in and around Ann Arbor, Mr. Rubbish. Private companies such as this one may be forced by their customers to offer a recycling opportunity, or they may simply want the positive publicity from offering it; but they do not want to lose much, if any, money on it. Accordingly, Mr. Rubbish requires extensive household separation, charges 35 cents per bag of recyclables, and collects only paper, metal, (#1 and #2) plastic, and alkaline batteries. For a long time, Mr. Rubbish accepted no glass containers, but *clear* glass is now accepted if separated (because collection with the rest of the trash and recyclables tends to break bottles) and unbroken, cleaned, and with cap and ring removed.

> *"In 2003, the U.S. electronics recycling industry was estimated to generate over $700 million annually and expected to grow to four or five times that by 2010."*

Electronics Recycling Is a Thriving, Environmentally Sound Business

Elizabeth Grossman

In this viewpoint taken from her book High Tech Trash, *environmental writer Elizabeth Grossman makes the case that recycling electronics is a large and growing industry. The growth in the volume of electronics waste has led to a new business of recycling this waste, she reports. The ideal use of this old equipment, she asserts, is refurbishment; repaired and updated equipment can be resold or donated to schools and nonprofit organizations. Materials and mining companies process the equipment that cannot be reused; the materials contained are a rich source of valuable metals. Both refurbishment and recycling help to reduce waste while employing people and making profits for companies, Grossman maintains.*

As you read, consider the following questions:

1. According to CompuMentor, as cited by the author, how much more energy efficient is refurbishing an older computer than manufacturing a new one?

2. According to Grossman, what are the two metals that companies are primarily interested in recycling from old circuit boards?

3. When did the large-scale recycling of electronics start, according to the author?

Researchers scrutinizing the environmental efficiency of high-tech equipment have an ongoing debate about the relative ecological merits of recycling and reuse. A study published in September 2004 by CompuMentor, a California-based nonprofit with an interest in reuse, found that refurbishing an old computer for reuse was twenty times more energy efficient than recycling. And by their estimate, only about 2 percent of used PCs find their way to a second-generation user. Eric Williams, who specializes in the life-cycle analysis of high-tech equipment, has calculated that reselling a computer or getting it to a second user saves 8.6 percent of the amount of energy required to make a new computer. Upgrading a computer, Williams estimates, saves 5.2 percent of that amount, and recycling a computer saves 4.3 percent of that energy.

Schools Benefit from Refurbished Electronics

Many manufacturers, including Dell, Hewlett-Packard, Intel, and Microsoft, all participate in programs that facilitate refurbishment and placement of used computers with schools and other nonprofits both at home and abroad, but reuse of high-tech equipment has yet to take off in the United States in a way that would even begin to put a dent in sales of new equipment. And it's almost harder to reuse or pass software on to a second or third user than it is hardware—if you follow the rules of the licensing agreement, that is, but organizations like CompuMentor are working on this and have a program with Microsoft that enables software to be transferred to qualifying second-users. Given the pace of high-tech innovation, until there are substantial design changes that allow new features to be added to older machinery, the refurbished, reused equipment will always lag behind. Although for many users, this may not be a hindrance

There are many admirable programs that make used computer equipment available to schools and nonprofits. Most are aimed at extending the life of relatively new equipment—often equipment from businesses, which tend to replace computers more often than individuals and do so in large numbers. There are nonprofits, like one I've visited in Portland, Oregon, called Free Geek, where donated equipment is refurbished or dismantled for parts that are then used to build new computers that are available either in trade for work at Free Geek or at low cost. And there are programs that send used computers to schools and other nonprofits in less well-off countries.

Most of these programs have inspiring stories that deserve to be told. But what makes these stories interesting and inspiring, I think, has more to do with the social roles they play in education and community development than with solving the fundamental problems associated with e-waste. That said, reuse is one of the best ways to extend the life of high-tech equipment and thereby reduce some of the production and consumption of new products that contributes to the ever-increasing piles of e-waste. Of course, at some point, second-hand computers will reach the end of their useful lives and need to be disposed of as well. Unless we make sure that happens properly—and that design and materials change to make recycling and disposing of high-tech electronics less problematic—we'll have simply slowed or shifted the e-waste stream.

And if any used computer is going to find its way into the recycling system, says Kevin Farnam, manager of environmental strategies and sustainability for Hewlett-Packard's corporate group in Houston, access must be convenient for both consumers and producers. "If it's not convenient, you've lost from the start," says Farnam.

Cyberage Mining

If I did send my computer off for recycling, one of the places it might end up is in the electronics recycling facility in Roseville, California, operated by Noranda—a Canadian mining company that's one of the world's major producers of zinc, nickel, and copper—and begun in partnership with Hewlett-Packard, the world's number one manufacturer of printers and their supplies and the world's second largest manufacturer of notebook computers. Why is one of the world's largest metals producers messing around with used computers, machines that contain more plastic than any other material?

"Printed circuit boards are probably the richest ore stream you're ever going to find," said Paul Galbraith of Concurrent Technologies Corporation to the audience of electronics recyclers at the EPR2 [Electronic Product Recovery and Recycling Project conference] in 2002. Mining companies like Noranda and the Swedish copper and zinc giant Boliden Mineral AB, known for extracting metals from the earth, are now mining circuit boards. Scott Pencer of Noranda calls this "above-ground mining." Instead of traveling the globe and making deep holes in the earth to extract their quarry, these companies are shredding and then melting and smelting circuit boards plucked out of old high-tech equipment to extract the valuable metals for resale and reuse. The individual quantities may be considerably smaller but, unlike prospecting for a new lode, a twenty-first-century circuit board miner knows what's going to be found—and the territory is a lot smaller to explore.

The metals of interest to companies like Noranda and Boliden are primarily copper and gold, but circuit boards also contain silver, platinum, and palladium, as well as some other nonprecious metals that can be recycled. Metals generally make up 30 to 50 percent of a circuit board, and while a typical sixty-pound desktop computer is only about 0.0016 percent gold, that gold is almost 100 percent recyclable. As a couple of newsletters I came across put it, "there's gold in them thar circuit boards and handsets!"

A New, Highly Specialized Industry

The outskirts of Roseville, about a twenty-five-minute drive northeast of Sacramento, are an architecturally unremarkable California locale. New shopping malls and suburban housing developments are in progress when I drive through in March of 2004. Surrounded by a chain-link fence, with some loading docks and parking spaces, and located just off a local highway, the 200,000-square-foot warehouse-like Noranda Recycling facility gives little, if any, clue from the outside as to what goes on within. From the area with the front offices where I'm given a hard hat and safety glasses, I'm guided into a room that seems large enough to house a small regional airport and is filled with what looks to be acres of computer equipment. I've entered, I realize, one of the places where these high-tech devices begin their journey to the electronics afterlife.

Like any highly specialized industry, electronics recycling is a world unto itself. Wander the floor of an electronics recyclers' conference and you'll come away with brochures that say things like "Turn Worthless Waste into a Valuable Commodity," "Certified Destruction," "Crush, Shred, Pulverize," "From Waste to Raw Material," "Do not waste IT—We re-

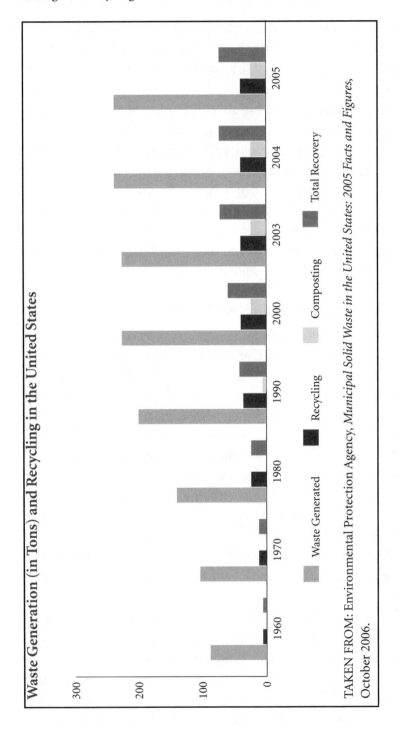

Waste Generation (in Tons) and Recycling in the United States

TAKEN FROM: Environmental Protection Agency, *Municipal Solid Waste in the United States: 2005 Facts and Figures,* October 2006.

cover it," "Shred-Tech," "Plastic Nation," and "ReCellular." You'll see ads from companies that make recycling equipment that feature phrases such as "Whole Lot of Shakin' Going On" and "Separation Anxiety?" Mark TenBrink, an operations manager for Noranda Recycling calls this the "tail end of the dog" of the high-tech electronics world.

Extracting Valuable Materials

"Typically we receive about 100,000 pounds a month but can receive 200,000 to 300,000 pounds," my guide Scott Sodenkamp, the operations manager at the Roseville facility, tells me as we walk the cement floor among the pallets laden with used computers, printers, fax machines, and photocopiers. "The most equipment we've ever gotten in at once is 400,000 pounds," he continues. This is just one of Noranda's recycling facilities. Others are located in San Jose, California, near Nashville, Tennessee, and in Canada. The Roseville facility, which has about one hundred employees, receives enough equipment for processing that it usually runs two eight-hour shifts a day, although it has the capacity to run around the clock.

Some of the recent arrivals at Noranda's Roseville facility are wrapped in plastic; some are bound with tape. Others are lying in huge cardboard cartons, and some are stacked on open metal shelves. A few forklifts, warning lights blinking and gentle horns beeping, are delivering boxes and pallets of equipment to various workstations. From somewhere out of sight comes the sound of heavy machinery. Despite the fact that this is a destination for discarded equipment, the place is extremely orderly.

Electronics recycling is a relatively new industry in the United States. The International Association of Electronics Recyclers (IAER) calls theirs an "emerging" industry, but a few mining and metals processing companies have been extracting valuable material from used equipment and putting it back into the metals market for several decades or more. Metals processors like Noranda, TenBrink tells me, began seeing electronics among their incoming scrap in the 1970s, but the finicky business of taking apart complex high-tech electronics equipment and separating out its many materials—so they can be turned into new sources of feedstock—only really got under way in the mid to late 1990s.

A Multimillion-Dollar Business

According to the IAER, there are from four hundred to five hundred electronics recyclers in the United States. Jerry Powell, editor at *E-Scrap News* had 950 North American e-scrap processors on his database—a number that does not include nonprofits, brokers, or reuse stores. And there are hundreds more high-tech electronics recyclers scattered around the world. This is not insignificant business. In 2003, the U.S. electronics recycling industry was estimated to generate over $700 million annually and expected to grow to four or five times that by 2010. Like companies in any other industry, electronics recyclers run from large and well established to small and struggling. Some are connected to corporations with operations that span the globe; some are fledgling family businesses. And many electronics recyclers specialize in a particular material, type of equipment, or component. To help guarantee a steady stream of material to keep the business running, many such companies have established relationships with high-tech equipment manufacturers. This access is crucial, because fundamental to all electronics recycling is processing enough "raw material" to turn back into a substantial and saleable quantity of feedstock.

Some electronics recycling facilities, like those operated by Noranda and Boliden, are associated with established metals-processing companies. "We had a big pile of stuff we didn't know what to do with. They knew what to do but didn't have the stuff," explains Renee St. Denis, manager of Hewlett-Packard's recycling programs, describing how HP and Noranda began working together.

Some electronics recyclers, like Singapore's Citiraya Industries, which had among its customers Intel, Nokia, and Hewlett-Packard, were founded by entrepreneurs who saw e-waste as a business about to boom. Others, like the Finnish company Kuusakoski, receive equipment through the Scandinavian electronics recycling consortium and send disassembled electronics to Boliden for materials recovery. Recyclers elsewhere in Europe and Japan work through similar arrangements with electronics manufacturers collaborating with metals companies, with regulatory and collection systems support provided by their governments.

The U.S. Environmental Protection Agency [EPA] does not currently license electronics recycling facilities, nor does any other federal, state, or local agency. Some states that have passed e-waste legislation are exploring adding special certification criteria for electronics recyclers to their existing health, safety, and environmental permits. The International Association of Electronics Recyclers has a certification program, but as of mid-2005 only a small number of recyclers had gone through the IAER process. The Institute of Scrap Recycling Industries and other industry groups are developing standards programs, and in 2005, in response to demand from those involved in e-waste issues, the EPA began discussing development of electronics recycling standards.

Periodical Bibliography

The following articles have been selected to supplement the diverse views presented in this chapter.

Martha Baer et al.	"The Recyclers," *Inc.*, November 2006.
Jon Birger and Paul Burall	"Can This Tiny Energy Company Really Change the World?" *Money*, July 1, 2003.
Sally Deneen	"How to Recycle Practically Anything," *E: The Environmental Magazine*, May/June 2006.
Matt Ewadinger and Scott Mouw	"Recycling Creates Jobs and Boosts Economy," *BioCycle*, October 2005.
Food Magazine	"Recycling Made Easy," September 10, 2004.
Anne Kandra et al.	"A Computer Is a Terrible Thing to Waste," *PC World*, January 2005.
Kivi Leroux Miller	"Is Recycling Disposable?" *American City & County*, May 2002.
Plastic News	"Recycling Success Requires Teamwork," May 22, 2006.
Bob Schildgen	"Advice for the Office, the Bathroom, and Beyond," *Sierra*, July/August 2005.
Neil Seldman	"The New Recycling Movement," *Waste to Wealth*, October 2003. www.ilsr.org/recycling/newmovement1.html.
John B. Stephenson	"Electronic Waste: Observations on the Role of the Federal Government in Encouraging Recycling and Reuse," *GAO Reports*, July 26, 2005.
Leland Teschler	"Save Energy, Don't Recycle," *Machine Design*, July 13, 2006.
Gareth Vans	"A Fashionable Fish Story," *BioCycle*, November 2004.

OPPOSING
VIEWPOINTS®
SERIES

CHAPTER 3

Do Specific Types of Waste Pose a Threat?

Chapter Preface

All waste is not created equal. Some categories of refuse generated by our society are more dangerous than others. As the business-oriented *Wall Street Journal* notes, "Dumps became awful Superfund cleanup messes not because of [consumer] trash but because of the improper disposal of industrial hazardous waste, a practice that has mostly stopped." While environmentalists would generally agree that there has been progress on some industrial wastes, they point out that there are other dangerous types of refuse that are still being generated or are newly emerging threats.

One long-standing waste problem is high-level radioactive waste. Largely the by-product of nuclear power generation—though also generated in the building of nuclear weapons—spent-fuel waste has been building up in the United States for the past half century. This refuse presents enormous technical problems; engineers must devise containers and facilities that can store deadly, radiation-generating spent fuel for tens of thousands of years.

Proposed solutions to the high-level radioactive waste problem illustrate a common split between what might be called the "environmentalist" versus the "technologist" mindset. Environmentalists generally view this waste as being too dangerous to handle, even with the best technology. They believe that nuclear power should be restricted until a permanent solution to the waste problem is found. The position of the Environmental Defense Fund is typical: "America must have a waste solution that not only has been vetted by the scientific community, but also is actually in place and working, before any expansion of the country's nuclear power generating capacity."

"Technologists" tend to believe that temporary solutions can be found to the nuclear waste question and that the future will bring a more permanent technological fix for the problem. "Extended temporary storage, perhaps even for as long as 100 years, should be an integral part of the disposal strategy. Among other benefits, it would take the pressure off government and industry to come up with a hasty disposal solution," write scientists John M. Deutch and Ernest J. Moniz in *Scientific American*. The hope is that technology will progress in those hundred years, making the final disposal of the waste easier.

The "environmentalist" versus "technologist" split is a fundamental area of disagreement that shapes authors' viewpoints on many differing aspects of garbage and recycling. In the following articles it can be found running through arguments that cover differing approaches to a wide range of categories of dangerous waste.

"*Given the sheer magnitude of e-waste generated each year, the problems that [their] toxins present increase exponentially as they progressively pollute the environment.*"

Society Must Address the Potential Dangers of E-waste

Morgan O'Rourke

Discarded personal computers and other electronic items, known collectively as "e-waste," have become a major source of toxic chemicals in landfills, according to Morgan O'Rourke in the following viewpoint. O'Rourke, managing editor of Risk Management *magazine, describes the potential harm of e-waste and outlines measures taken to combat the problem. Spurred on by new laws or the threat of legislation, some manufacturers are instituting recovery and recycling programs, he reports. These can be costly, however, and there is controversy, O'Rourke claims, over who should pay for such programs—the government, the consumer, or electronics producers.*

As you read, consider the following questions:

1. According to O'Rourke, what is the most prominent pollution danger from discarded cathode ray tubes?

2. According to the National Safety Council, as cited by the author, in the year 2001 what percentage of personal computers were recycled?

3. Under the European Union's Extended Producer Responsibility law, who has primary responsibility for e-waste generated by personal electronics, according to O'Rourke?

The vast amount of computers, televisions, mobile phones and the like that are disposed of every year all contain a variety of toxic substances. When electronics are dumped in landfills, these substances can leach into the soil and groundwater (regardless of whether the landfill is sealed or not). When the waste is incinerated, contaminants and toxic chemicals are generated and released into the air. Given the sheer magnitude of e-waste generated each year, the problems that these toxins present increase exponentially as they progressively pollute the environment and threaten to enter the food chain.

Heavy Metals

One of the most serious concerns is the cathode ray tubes (CRTs) used for computer monitors and television screens. California already prohibits CRT disposal in municipal solid waste landfills, and improper CRT disposal in Massachusetts can carry a fine of up to $525,000 per instance. This is because CRTs contain a wide range of toxic substances, the most harmful of which is lead, which can cause vomiting, diarrhea, convulsions, coma or even death at high levels of exposure. It can also damage the nervous system and kidneys, cause anemia and promote an increased risk of several types of cancer. Since a typical computer display contains an average of four pounds of lead, more than 250 million obsolete computers by 2005 would result in over one billion pounds of lead to consider.

Unfortunately, lead is far from the only toxin in e-waste that raises concern. Cadmium, which has a serious impact on kidney function, can be found in CRTs, batteries, resistors and housings. Mercury is highly concentrated in electronic components, batteries, switches and printed wiring boards and can cause brain and liver damage. In addition, harmful metals such as chromium, beryllium, barium and arsenic, as well as brominated flame retardants in plastics can be present in landfill leachate.

Another typical disposal scenario involves incineration. But like land-filling, this also exposes the environment to dangerous contaminants by generating and releasing these substances into the atmosphere, sometimes through complex chemical reactions. For example, the copper in printed circuit boards and cables is a catalyst for dioxin and furan formation when flame retardants are burned. Dioxins, which are carcinogenic and can cause immune and reproductive system damage, are also formed when housings and cables containing PVC are incinerated. The resulting fly ash from incinerators is a large source of heavy metal contamination, as well.

Recycling and Producer Responsibility

In order to combat the environmental impact of improper electronic waste disposal, many organizations have opted to recycle their old technology. But while recycling is growing in popularity, rates are still low. In 2001, the National Safety Council reported that only 11% of personal computers retired in the United States were recycled. But with increased attention being given to technology recycling through government initiatives, manufacturers' programs and environmental group campaigns, this number is rising.

For instance, the Environmental Protection Agency's [EPA's] Resource Conservation Challenge seeks to increase the national recycling rate and reduce fine environmental impact of electronic products during their production, use and disposal. . . .

Many computer manufacturers have also begun to address the electronic waste problem with their own end-of-life management programs. Companies usually offer a combination of recycling, trade-in, take-back or leasing programs that take the responsibility of disposing of old electronics away from the purchaser. . . .

While these and other manufacturer-based recycling efforts are commendable, it is likely that for some, a sense of environmental responsibility was not the only impetus to their development. In 2001, the European Union adopted a system called Extended Producer Responsibility that requires electronics manufacturers to take full responsibility for the recovery and recycling of e-waste as well as begin to phase out the use of hazardous materials altogether. Meanwhile in the United States, regulations have been passed on the state level to establish responsibility guidelines for the management of e-waste. In May, for example, Maine passed a similar law requiring manufacturers to take the primary responsibility for managing discarded products by establishing effective collection and recycling programs.

The problem for some manufacturers, however, is that recycling efforts are costly and the legislative trend towards producer responsibility that started in Europe and has now appeared in the United States places the financial burden squarely on the shoulders of the manufacturer. They fear that it will increase costs and put companies at a competitive disadvantage from state to state. Manufacturers also do not generally approve of California's system of mandatory consumer-paid advance recovery and recycling fees because it unfairly shifts the cost onto customers. "If we just do that," says

---

Electronic Waste Facts

- More than two million tons of e-waste end up in landfills every year.

- By 2005, 130 million mobile phones [were] discarded annually, accounting for 65,000 tons of waste.

- The average lifespan of a new computer has decreased from 4.5 years in 1992 to an estimated two years in 2005.

- The average consumer has two to three computers in storage in garages, closets, and other areas.

> *U.S. Environmental Protection Agency and the National Safety Council, October 2004.*

[Hewlett-Packard executive David] Lear, "the burden goes to our customer, we have no incentive to design our product any differently, and we have no incentive to bring these products back ourselves and reuse the materials."

Sharing the Burden

For the most part, manufacturers support responsible electronic waste management but believe that the responsibility should be shared between manufacturers, government, and private waste and recycling companies to achieve maximum efficiency.

Environmental groups like the Silicon Valley Toxics Coalition (SVTC) are pushing for producer responsibility as well, although they do not necessarily favor shared responsibility. Sheila Davis, director of SVTC's Clean Computer Campaign, says that if companies designed more environmentally sound products in the first place, the cost of recycling would be reduced. And if manufacturers can use the recycled material in new products, they will also reduce their raw material expenses. For the SVTC, the primary focus of the Maine law is not simply that it requires manufacturers to focus on recycling. "Recycling is a big part," says Davis. "But if you don't design the product so that it can be recycled, it won't be recycled in a cost-effective manner. Our goal is to have the manufacturers redesign their products so that they are more environmentally sound."

So increased pressure from government and environmental organizations has led many manufacturers to proactively address the e-waste problem, and since these efforts are well received by customers, it bolsters a company's reputation in the marketplace.

Hewlett-Packard and Dell, for example, both signed a producer responsibility statement drafted by the Computer Take Back Campaign, an organization that promotes producer responsibility as the optimal way to eliminate hazardous electronic waste. The Campaign is made up of various environmental organizations, including the SVTC, the Grass Roots Recycling Network, and the Texas Campaign for the Environment. The statement reads in part, "We support the policy of producer responsibility in the U.S. for electronic products at the end of their useful lives, wherein brand-name manufacturers/producers work with consumers and state and local governments to properly collect and manage electronic products in an environmentally responsible fashion. Manufacturers and producers accept responsibility for continually improving the environmental aspects of the design of their products and for the end-of-life management of their products." The companies' move towards what the Campaign believes is a comprehensive and sustainable solution to the e-waste problem helped earn them top marks on the Campaign's annual Computer Report Card, which grades the environmental efforts of various manufacturers.

The Penalties for Hazardous Waste

As manufacturers continue to develop recycling and environmental responsibility policies, end users have more options when considering what to do with their outdated technology. But companies need to be aware that it is important that the materials be disposed of properly.

As a result of environmental concerns, most electronic waste is classified as hazardous or universal waste and therefore subject to a host of regulations that have been established to prevent the improper disposal of toxic materials.

Through the Comprehensive Environmental Response, Compensation and Liability Act (CERCLA), commonly known as Superfund, and the Superfund Amendments and Reauthorization Act (SARA), the EPA identifies and remediates hazardous waste sites and can assess significant fines and penalties to the companies responsible for the contamination. . . .

Even Recycling Can Be Risky

Even recycling poses its own hazards if not done properly. According to the Computer Take Back Campaign, while some discarded equipment is handled by firms that operate under strict environmental controls and high worker safety protections, many other firms do not operate under these controls, and simply remove the valuable materials from the equipment and send the remaining scrap to landfills or incinerators. Without adequate protections, workers dismantling discarded electronic equipment are exposed to many chemical compounds that can cause a host of negative health effects.

It is also suspected that considerably more equipment is shipped to China and other Asian nations, where it is dismantled under unsafe conditions, poisoning the local people, land, air and water.

"There is no scientific evidence that substances from e-waste present a discernable risk to human health or the environment when disposed of in municipal landfills."

The Dangers of E-waste Are Exaggerated

Dana Joel Gattuso

In the following viewpoint environmental analyst Dana Joel Gattuso claims that the dangers from electronics waste, or e-waste, are exaggerated. He relies on studies that have shown that consumer electronics in landfills contribute to only slight increases in lead in surrounding soils and groundwater. The levels are well within the range considered safe for human health by the United States government, he argues.

As you read, consider the following questions:

1. According to Gattuso, what percentage of the municipal solid waste stream was made up of discarded consumer electronics in 2001?

2. What substance constitutes "almost all" of the heavy metal in landfills, according to the author?

3. According to Gattuso, how much per ton does it cost to recycle e-waste and to landfill e-waste?

In the home, computers are becoming as commonplace as toasters. Rapid improvements in technology and design, as well as increased competition, have made home computers more affordable than ever. [In 1991], 16 percent of U.S. households owned a home computer; today, more than half own at least one computer.

Old Computers in Attics

Innovation and affordability have also enabled computer manufacturers to roll out new, faster, and upgraded models at a prodigious rate. Since 1981, more than a billion personal computers have been sold worldwide—400 million of those in the United States. In 2003 alone, more than 50 million computers were sold in the U.S.

A natural by-product of the home computer revolution is the growing number of outdated computers. Between 1997 and 2003, there were an estimated 254 million obsolete computers in the U.S. Projections show another 250 million will become obsolete between 2004 and 2007, though the annual number of outdated machines is expected to level off at around 63 million by 2005, according to the National Safety Council.

What is the fate of the used home computer in the U.S.? Most—an estimated 75 percent—are believed to be stockpiled in people's homes, typically in basements, attics, or garages. Fourteen percent are recycled or reused. And, surprisingly, only 11 percent are buried in landfills.

Misperceptions Fuel Fear about E-waste

Concern over the rapid growth of used computers and what to do with them once they expire has placed the issue of how best to handle electronic waste—or "e-waste"—at the forefront of waste policy at the federal, state, and local levels. Increasingly, propaganda fueled by politically driven environmental activists and a misinformed media is turning concern into hysteria. Fears are largely based on the following myths:

Electronic waste is growing at a rapid and uncontrollable rate and is the fastest growing portion of the municipal waste stream. While the amount of e-waste has been increasing, it remains a tiny percent of the total municipal solid waste stream. According to the Environmental Protection Agency (EPA), e-waste—including discarded TVs, VCRs, DVD players, and audio systems, as well as personal computers, fax machines, and printers—constituted only 1 percent of the total municipal solid waste stream in 1999, the first year EPA calculated electronics discards. Data for 2001 again showed electronic devices had not increased as a percent of total municipal waste but remained at 1 percent.

Nor is e-waste growing at a rapid rate. National Safety Council (NSC) data show that the number of discarded computers will . . . begin to decline [after 2005]. While improved technology can quickly make machines obsolete, it can also extend the lifespan of the next generation of computers. More powerful microchips will soon provide machines with much greater capacity.

Computers buried in landfills endanger public health because they contain toxic materials such as lead, cadmium, and mercury that can leak out into the soil and groundwater. Cathode ray tubes (CRTs), the most common type of computer display monitor, typically contain four pounds of lead to protect users from the tubes' x-rays, the same way a lead vest protects patients who have x-rays. Because lead is a health risk at high exposure levels, many lawmakers are rushing to ban display monitors and other electronics from municipal landfills, fearing that the lead and other toxic metals can leak out into the ground soil. Overwhelmingly, lawmakers and the popular press point to the work of Timothy Townsend, Associate Pro-

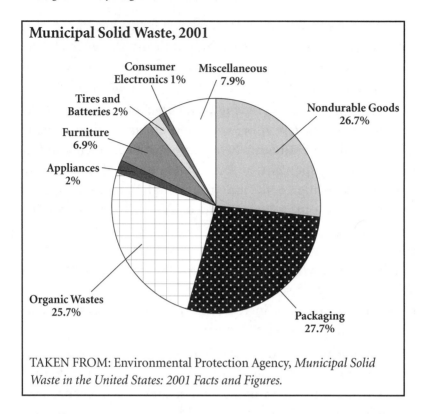

Municipal Solid Waste, 2001

Consumer Electronics 1%

Miscellaneous 7.9%

Tires and Batteries 2%

Nondurable Goods 26.7%

Furniture 6.9%

Appliances 2%

Organic Wastes 25.7%

Packaging 27.7%

TAKEN FROM: Environmental Protection Agency, *Municipal Solid Waste in the United States: 2001 Facts and Figures.*

Consumer electronic waste, including computers, made up only a small percentage of all solid waste disposed of in 2001.

fessor of Environmental Engineering Sciences at the University of Florida and a leading expert on solid waste, who has been studying for over six years the potential for lead to leak out—or "leach"—from computer monitors, TVs, and other electronic components into the ground soil.

Media Ignores Important Research

But incredibly, the media has only reported on Townsend's earlier research using the questionable Toxicity Characteristic Leaching Procedure [TCLP] test, a method used by the EPA that attempts to simulate the conditions of a landfill under a worst-case lab test by soaking tiny samples of e-waste in an acid solution and testing their levels of toxicity. Townsend himself concluded in his 1999 report that although his tests showed that 21 of 30 color monitors failed the EPA-defined regulatory limit, the EPA's leaching procedure tests do not mimic what actually occurs in landfills, and "the authors do not attempt to draw conclusions beyond [the specific results of the lab test] in regard to the implications of the lead leaching from CRTs." He also wrote: "The fact that the [EPA's] TCLP test may not represent the true condition of CRTs upon disposal was not an issue of discussion in this research;" and "the leachate concentrations measured [by the EPA lab test method] may not accurately reflect the concentrations observed under typical landfill conditions." Yet newspaper write-ups following the study's release reported only the dangers of lead from electronic monitors in landfills and triggered a panic among many policymakers and lawmakers calling for a ban of all CRTs from landfills.

Recognizing the EPA test's potential inadequacy in replicating landfill conditions, Townsend and his colleague Yong-Chul Jang conducted a new test in 2003, using 11 actual landfills containing electronic waste and other municipal waste and debris. Specifically, they tested soil from landfills containing waste from color TV and computer monitors, shown in his previous EPA lab tests to leach the highest levels of lead. They also tested soil containing waste from home computer circuit boards, which also contain lead. Comparing the landfills' concentrations of heavy metals in the ground-soil waste—called "leachate"—with levels from the earlier EPA lab test, Townsend found concentrations of lead from the landfilled computer monitor leachate to average only 4.1 milligrams per liter (mg/L)—that's less than 1 percent of what the laboratory studies suggested would be the case (lab tests suggested the monitors would leak 413 mg/L of lead in leachate).

Similarly, he found only 2.2 mg/L [milligrams per liter] of lead in landfill leachate from computer circuit boards—a little more than one percent of the 162 mg estimated in lab tests.

Hence, it is highly likely that actual landfill releases of these heavy metals are far lower than EPA estimates. These differences are far from minimal. As Townsend concludes: "For those state and local governmental agencies wrestling with whether to ban discarded electronics from landfills, the results of this work suggest that lead leaching from [computer circuit boards] and [TV and computer monitors] will be less than might be estimated using EPA's TCLP results." Even more importantly, concentrations from his landfill samples were comfortably below EPA's standards of 5.0 mg/L. Yet it is important to note that these materials would not even enter the environment, since landfill operators collect and dispose of [them] in a safe manner.

Additional Studies

Townsend is further researching leachate and waste settlement from actual landfills. His . . . study . . . completed [in 2005] involves constructing landfills and filling them with simulated municipal solid waste containing e-waste. But on the overall question of whether e-waste leaches in landfills, Townsend says "there is no compelling evidence."

Other recent studies confirm that lead and other metals contained in landfills are safely contained. A year-long study by the Solid Waste Association of North America (SWANA) Applied Research Foundation, released in March 2004, concludes that heavy toxic metals, including lead, do not pose an existing or future health threat in municipal solid waste landfills. The foundation reviewed existing research and concluded that landfills' natural conditions, such as precipitation and absorption, provide chemical reactions and interactions that prevent heavy metals from dissolving into the soil. They concluded that out of 130,200 tons of heavy metals placed in municipal landfills in 2000 from electronics, batteries, thermometers, and pigments, almost all—98 percent—was lead. Cadmium and mercury made up the remaining amount. According to the authors, "The study presents extensive data that show that heavy metal concentrations in leachate and landfill gas are generally far below the limits that have been established to protect human health and the environment." The report was peer reviewed by an independent panel of researchers in the field, including Timothy Townsend. Oddly, neither this report nor Townsend's recent research comparing EPA lab tests' leachate with actual landfill leachate was ever reported by the general media.

Even if the natural conditions that prevent leaching did not occur, the sophisticated engineering and monitoring of today's modern municipal landfills, governed by stringent state and federal regulations and performance standards, prevents lead and other heavy metals from leaching. MSW [municipal solid waste] landfills are constructed with thick layers of clay and thick, puncture-resistant liners that keep waste from coming into contact with soil and groundwater. Also, landfills today are constructed with a leachate collection system—a system of pipes that carries any excess leachate out of the landfill and into a separate leachate collection pond where it is then tested and treated. In addition, landfills are surrounded by groundwater-monitoring stations which capture samples of groundwater and continuously test for any possible leaks.

No Scientific Evidence of Risk from Leachate

In summary, there is no scientific evidence that substances from e-waste present a discernable risk to human health or the environment when disposed of in municipal landfills. Yet widespread fear that lead and other metals in landfills can leach and present a health hazard has provoked lawmakers in a handful of states—California, Maine, Massachusetts, and Minnesota—to ban desktop display monitors from landfills; another half a dozen have pending legislation.

Ironically, the problem is not so much electronic waste it-self, but what to do with the enormous quantities of e-waste if lawmakers choose to ban it from landfills. Furthermore, lead and other compounds are considered by some experts to be safer when contained in landfills than during the recycling process when they become exposed. Finally, the cost difference is astronomical. Where a ton of e-waste can cost $500 to re-cycle, it costs only $40 to landfill.

| "Here's the twist: with nuclear waste,
procrastination may actually pay."

Short-Term Storage of Spent Nuclear Fuel Is Safe and Effective

Matthew L. Wald

In the following viewpoint, longtime technology writer Matthew L. Wald maintains that the federal government's project to store high-level radioactive waste at Yucca Mountain, Nevada, has been a failure for both technological and political reasons. Wald believes a central above-ground facility would be better. During the 100- to 150-year lifespan of such a facility, according to Wald, new technology and the demand for recycled nuclear fuel may make the problem of truly permanent storage of nuclear waste easier to solve.

As you read, consider the following questions:

1. In Wald's view, what are two reasons that temporary storage, lasting perhaps a hundred years, might be better than rushing waste into permanent, buried storage?

2. How might the development of ceramics technology help with nuclear waste storage, in the author's opinion?

Matthew L. Wald, "A New Vision for Nuclear Waste," *Technology Review*, vol. 107, December 2004, pp. 38–44. © 2004 Technology Review, Inc. Reproduced by permission.

3. Why isn't it economical to reprocess (recycle) nuclear
   fuel right now, according to Wald?

When American Airlines Flight 11 flew at low altitude down the Hudson River valley on the morning of Sept. 11, 2001, its target was the north tower of the World Trade Center. But its impact is still being felt at a cluster of buildings it passed about five minutes before it reached lower Manhattan, at a nuclear-reactor complex called Indian Point in Buchanan, N.Y. Adjacent to the site's two operating reactors are two buildings packed with highly radioactive spent-fuel rods, in pools of water 12 meters deep and tinged Ty-D-Bol blue by boron added to tamp down nuclear chain reactions. The soothing hum of the pumps that circulate the building's warm, moist air—and, critically, keep the water cool—lends an atmosphere of industrial tranquility.

Without that cooling water, the fuel cladding might overheat, melt, catch fire, and release radiation. Whether the impact of a Boeing 767 like Flight 11 could drain one of the pools and disable backup water pumps, starting such a fire, is far from clear. Nevertheless, the threat of terrorism in general and the flyover of Flight 11 in particular have reignited the debate about why all of this dangerous fuel is still here— indeed, why all spent fuel produced at Indian Point in three decades is still here—and not at Yucca Mountain [Nevada], the federal government's burial spot near Las Vegas, where it was supposed to be shipped beginning [in 1998].

Building Temporary Storage

[In 2004], a construction project began at Indian Point that will allow the fuel to be pulled out of the pools. But it's not going to Yucca. The government says Yucca won't be ready until 2010. Executives in the nuclear industry say a more likely date is between 2015 and never. So instead of traveling to Nevada, Indian Point's fuel is traveling about 100 meters, to a bluff overlooking the Hudson River. On a late-summer day [in 2004], a backhoe tore out maple and black-walnut trees to make way for a concrete pad. Beginning [in 2005], the first of a planned 72 six-meter-tall concrete-and-steel casks will be placed there, a configuration that adds storage capacity and thus allows the twin power plants to keep operating. Though they provide a hedge against a worst-case fuel-pool meltdown, these casks are merely another temporary solution. The fact that they're needed at all represents the colossal failure of the U.S. Department of Energy's Yucca plans and technology.

Yet as engineering and policy failures go, this one has a silver lining. Conventional thinking holds that Yucca's problems must be solved quickly so that nuclear waste can be squirreled away safely and permanently, deep within a remote mountain. But here's the twist: with nuclear waste, procrastination may actually pay. The construction of cask fields presents a chance to rethink the conventional. The passage of several decades while the waste sits in casks could be immensely helpful. A century would give the United States time to observe progress on waste storage in other countries. In the meantime, natural radioactive decay would make the waste cooler and thus easier to deal with. What's more, technological advances over the next century might yield better long-term storage methods. "If it goes on for another 50 years, it doesn't matter. It could go on for 100 or 200 years, and it's probably for the better," says Allison Macfarlane, a geologist at MIT and coeditor of a forthcoming book on Yucca. "We've got plenty of time to play with it." . . .

Cooler Fuel

The argument against casks is that they are merely temporary, not meant to serve longer than perhaps 100 years, and that they are a kind of surrender, leaving this generation's waste problem to a future generation to solve. Yet their impermanence is exactly what's good about them. A century hence, spent reactor fuel will be cooler and more amenable to permanent disposal. In fact, within a few decades, the average fuel bundle's heat output will be down to two or three hair dryers. After 150 years, only one-thirty-second of the cesium and strontium will remain. The remaining material can be buried closer together without boiling underground water. Reduced heat means reduced uncertainty.

Granted, spent fuel will be far from safe after such a relatively short period. Even after 100 years, it will still be so radioactive that a few minutes of direct exposure will be lethal. "It's many, many, many thousands of years before it's a no nevermind," says Geoffrey Schwartz, the cask manager for Indian Point, which is owned by Entergy Nuclear. "But the spent fuel does become more benign as time goes by."

The fuel could be more valuable, too. For decades, industry and government officials have recognized that "spent" reactor fuel contains a large amount of unused uranium, as well as another very good reactor fuel, plutonium, which is produced as a by-product of running the reactor. Both can be readily extracted, although right now the price of new uranium is so low, and the cost of extraction so high, that reprocessing spent fuel is not practical. And the political climate does not favor a technology that makes potential bomb fuel—plutonium—an item of international commerce. But things might be different in 100 years. For starters, the same fuel could be reprocessed much more easily, since the potentially valuable components will be in a matrix of material that is not so intensely radioactive.

Future Recycling Opportunities

And in 100 years, advances in reprocessing technology might make the economics compelling. The existing American technology dates from the Cold War and involves elaborate chemical steps that create vast quantities of liquid waste. But an alternative exists: electrometallurgical reprocessing. Though research into the technique has lagged of late because of the economic climate, the concept might be taken more seriously in the future. Electrodes could sort out the garbage (the atoms formed when uranium is split) from the usable uranium (the uranium-235 still available for fission and the uranium-238 that can be turned into plutonium in a reactor), in something like the way jewelers use electrometallurgy to apply silver plate. Resulting waste volumes would be far smaller.

Perhaps most importantly, in 100 years, energy supply and demand might be very different. Reprocessed nuclear fuel might well become a critical part of the energy supply, if the world has run out of cheap oil and we decide that burning coal is too damaging to our atmosphere. If that happens, we might have 1,000 nuclear reactors. On the other hand, we might have no reactors, depending on the progress of alternate energy sources like solar and wind. At this point, it's hard to tell, but we are not required to make the decision now; we can put the spent fuel in casks for 50 years and then decide if it is wheat or chaff.

There is a final, more practical reason that we might choose to take the plutonium out of spent fuel for reactor use: it makes the remainder easier to store. For the most part, what's left will not be radioactive for nearly as long, and the sheer volume of material will be lower. Mark Deinert, a physicist at Cornell University, says reprocessing, like recycling, removes about half of the material from the waste, dramatically decreasing storage costs and effectively doubling the capacity of a facility like Yucca.

Betting on Better Storage

While nuclear waste would be easier to handle in 50 or 100 years, it would still require isolation for several hundred thousand years. But there is every reason to expect that storage technology will improve in the next century. When we decide to permanently dispose of the waste, either after reprocessing

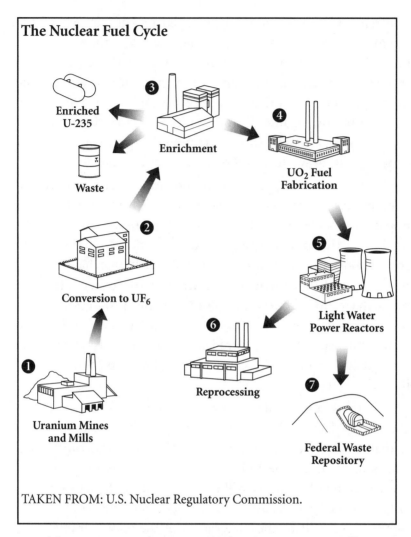

The Nuclear Fuel Cycle

❸ Enrichment

Enriched U-235

Waste

❹ UO$_2$ Fuel Fabrication

❷ Conversion to UF$_6$

❺ Light Water Power Reactors

❻ Reprocessing

❶ Uranium Mines and Mills

❼ Federal Waste Repository

TAKEN FROM: U.S. Nuclear Regulatory Commission.

or without reprocessing, we may be smarter at metallurgy, geology, and geochemistry than we are now.

Today, the basic technology at Yucca is a stainless-steel material called alloy 22, covered with an umbrella of titanium—a "drip shield" against water percolating down through the tunnel roof. That could look as primitive in 100 years as the Wright brothers' 1903 Flyer looks to us [today]. Or it might simply be obsolete. Space-launch technology could become as reliable as jet airplanes are today, giving us a nearly foolproof way to throw waste into solar orbit. The mysteries of geochemistry might be as transparent as the human genetic code is becoming, which would mean we could say with confidence what kind of package would keep the waste encased for the next few hundred thousand years.

Or there might be easier ways to process the waste. For example, particle accelerators, routinely used to make medical isotopes, could provide a means to make the waste more benign. The principle has already been demonstrated experimentally: firing subatomic particles at high-level radioactive waste can change long-lived radioactive materials to short-lived ones. Richard A. Meserve, a former chairman of the U.S. Nuclear Regulatory Commission and now the chairman of a National Academy of Sciences panel on nuclear waste, says this technology, known as transmutation, might become more practical in 100 years. The technology of accelerators has advanced in the last few years, he says, and it is a good bet that it will continue to do so.

Safer Technologies May Be Close at Hand

Some alternative storage technologies may need only a few more years of research and development. One is ceramic packaging. Ceramics have good resistance to radiation and heat, and they don't rust. At the moment, nobody casts ceramics big enough to hold fuel assemblies, which are typically about four meters long. But there is no theoretical limit to the sizes of ceramics; there has simply been no economic incentive to make giant ones. Nor will there be, until the only likely customer for them, the Energy Department, decides that the metal it is shopping for now isn't up to the job.

Another alternative calls for mixing waste with ceramics or minerals to form a rocklike material comprising about 20 percent waste. The waste would be chemically bound up in stable materials that are not prone to react with water. With a few decades' grace time, engineers could build samples and test them in harsh environments. But even though the idea has been around for more than 10 years, no one has put serious research money into it, since its only possible American customer, the Energy Department, has been committed to Yucca.

| "The state of Nevada ... makes its case against Yucca Mountain on a technical basis; there is little discussion of the issues of equity and regional politics, though these are central to Nevada's objections."

Technical and Political Difficulties Prevent a Permanent Nuclear Waste Storage Solution

Allison Macfarlane

Allison Macfarlane is a professor at George Mason University and author of a book, Uncertainty Underground, *about the Yucca Mountain, Nevada, nuclear waste disposal facility. In this viewpoint, Macfarlane highlights the policy problems inherent in building a storage site which must contain radioactive waste for tens of thousands of years. Scientific models predict that the project is safe, but there are many variables which cannot be foreseen. Supporters of the project claim the models are 'scientific proof' of Yucca Mountain's safety, opponents point to unforesee-able problems. However, the amount of nuclear waste grows daily; policy makers will have to decide soon.*

As you read, consider the following questions:

1. How long has the United States' high-level radioactive waste been building up in temporary storage

2. What are the two primary natural factors that could cause the Yucca Mountain facility to fail?

3. If Yucca Mountain is operational will the United States' problem with nuclear waste be solved? Why or why not?

The United States has been accumulating high-level nuclear waste since the dawn of the atomic age. The nuclear fuel cycle remains an incomplete circle; a process for the final step, disposal, has never been resolved. Today, 60 years' worth of radioactive wastes have piled up across the country at more than 120 sites—all of which are supposed to be temporary. By the end of 2005, about 55,000 metric tons of spent nuclear fuel had built up at reactor sites around the country, plus about 15,000 metric tons of high-level waste from the nuclear weapons complex. For nearly a quarter-century, the government has focused its efforts to find a permanent place to store its highly radioactive waste on the idea of deep, underground geologic repositories—in particular, a site at Yucca Mountain, Nevada. But billions of dollars and many years later, the United States is still struggling to find a solution to its literally grow-ing problem of nuclear waste.

It may seem today that there has only ever been one seri-ous contender for a permanent repository—Yucca Mountain. But that is not the case. Before Yucca, a number of other sites were considered.

It was almost 50 years ago, in 1957, that the National Academy of Sciences first framed a solution to the high-level nuclear waste problem, suggesting that it should be emplaced in a geologic repository. Little action was taken until 1970, when the Atomic Energy Commission (AEC) selected an old salt mine in Lyons, Kansas, as a repository. It was to be the nation's first lesson in siting struggles. The AEC had not investigated the location well enough, and it turned out that the Lyons mine was far from watertight, thanks to an adjacent mine and old oil and gas drill holes. (Moving water can transport nuclear waste to the environment.) The Lyons project turned into a public relations disaster, and the resulting backlash swung the momentum away from underground repositories for a short while, toward the use of long-term, above-ground storage canisters.

Just a few years later, in 1974, India detonated a nuclear test device that used plutonium diverted from its Cirus research reactor. This marked a turning point for the back-end of the U.S. nuclear fuel cycle. Until that point, the U.S. nuclear industry had been planning to reprocess spent fuel, using the plutonium extracted from it to fuel a new fleet of fast breeder reactors. In response to the proliferation dangers that India's detonation revealed, the Ford administration "indefinitely deferred" the reprocessing of spent fuel. President Jimmy Carter continued the prohibition, hoping to set an example for other countries and thereby avoid the potential for diversion of plutonium into nuclear weapons.

Nuclear reactors were designed to store only a minimal amount of spent fuel in their onsite storage pools, where used fuel was to cool down before being sent to reprocessing facilities. The reprocessing ban meant that spent fuel would not be removed from reactor sites and would soon overwhelm their onsite storage capacity. Congress reacted to this predicament by passing the 1982 Nuclear Waste Policy Act (NWPA). The act established geologic repositories as the long-term solution to the problem of storing high-level nuclear waste, and it set in motion the process to site and develop such repositories. The Energy Department was tasked with identifying sites and evaluating them, and the Environmental Protection Agency (EPA) was to developed standards that the Nuclear Regulatory Commission (NRC) would use to determine whether a site should be licensed.

The act, which required the federal government to open a permanent repository by January 31, 1998, envisioned a minimum of two national storage sites. It was understood that if one were in the Western United States, the other would be located in the East. For the first repository, three locations were to be selected from a larger list and then characterized simultaneously, compared, and the best site selected. By 1986, Energy had selected three sites: Yucca Mountain, 90 miles northwest of Las Vegas; Deaf Smith County in the Texas panhandle; and the Hanford nuclear reservation in Washington State.

When Energy "indefinitely deferred" the search for a second site in 1986 (apparently due to a decreased need for a second site), a heated controversy erupted concerning the siting process. Texas, Washington, and Nevada all feared that the bulk of nuclear waste would land in their states. Congress cut off appropriations for development of a geologic repository, and the impasse was broken by the 1987 Nuclear Waste Policy Amendments Act. The legislation reduced the number of sites to be studied down to one: Yucca Mountain. The law also delayed a decision on the need for a second repository until 2010 and prohibited research on crystalline rock repositories at the behest of East Coast states that wanted to take themselves out of the running. As a result, much of the East Coast was eliminated from contention for a potential second site.

Not surprisingly, Nevada found the process established by the 1987 amendments unfair. Opponents of the site felt it singled out a politically weak state (both of Nevada's senators were junior) with no nuclear power plants to be a dumping ground for some of the nation's most dangerous waste. They have been fighting the repository at Yucca Mountain ever since.

By the early 1990s, it had become clear that Energy would not meet the January 1998 deadline to open a repository. Congress demanded that Energy at least prove that Yucca Mountain was a workable site, resulting in a viability assessment in 1998. At the same time, nuclear power plant owners, whose spent fuel pools were filling up, filed lawsuits against Energy for breach of contract because the department failed to take their waste. Energy continues to negotiate with some of the utility companies over their contracts. The NRC also approved nuclear reactors to use aboveground, dry casks as temporary onsite storage if their spent fuel pools were full. More than half the spent fuel pools at U.S. nuclear reactors are now at their capacity, and according to Energy Department figures, in less than 10 years, nearly every reactor will require dry storage.

In February 2002, Energy declared the Yucca Mountain site suitable for a geologic repository. By law, the state of Nevada was allowed a veto of the site, and in April 2002, the governor of Nevada declared his disapproval. But three months later, Congress overrode this veto and approved Yucca Mountain. Although the site has received presidential and congressional seals of approval, Energy is far from getting the final green light for actually building the repository, and it is still not clear whether Yucca is a viable long-term site.

Officially, Energy had only 90 days from congressional approval to submit a construction license application to the NRC, but Energy has not yet done so. Once Energy does submit the license application, the NRC has up to four years to review it and determine whether to grant the license. After that, Energy must submit a second license application in order to receive waste, and again, the NRC decides whether to approve it.

Why is it taking so long for Energy to turn in its license application for construction? Of many contributing factors, three major problems stand out. In July 2004, a federal appeals court rejected the EPA's radiation dose standard for Yucca Mountain, demanding that it evaluate radiation dose limits for a period of up to 1 million years; EPA's previous standard took into account only 10,000 years. Another serious delay was caused by the revelation last year of charges of scientific fraud at Yucca Mountain. Energy discovered e-mail messages dating from the late 1990s that suggested that scientists who worked on water transport issues at Yucca Mountain had falsified data to satisfy Energy quality assurance requirements. (Energy recently released a report stating that the falsifications did not impact the scientific integrity of its assessment of the site.) The third major difficulty is Energy's apparent inability to publish on its website within six months of submitting a license application, all supporting documentation for the Yucca Mountain site—as per NRC rules. Energy had originally intended to submit a license application by the end of 2004, but the NRC ruled it could not unless it made public all the supporting information.

Energy has developed a draft license application and recently announced that it will submit it sometime in the next few years; it promises to publish a schedule by this summer. Nevada has requested a copy of the draft application so that it may review Energy's case for the suitability of the Yucca Mountain site—a request the Energy has repeatedly denied. In February 2006, in response to a complaint from the state of Nevada, the NRC ruled in favor of Energy—overruling its own advisory board.

Active opposition to the site within Nevada also remains a thorn in the side of progress on the repository. The state has developed a multipronged approach to derailing the Yucca Mountain site, using courtrooms and Congress as its battlefields. Nevada continues to submit lawsuits on various aspects of nuclear waste disposal, including one against the NRC over the licensing process and against Energy over the siting of a railroad that would carry waste to Yucca Mountain. Nevada's

Democratic Sen. Harry Reid introduced a bill in December 2005, the Spent Nuclear Fuel On-Site Storage Security Act, which would allow Energy to take title to spent fuel while it remained in dry cask storage at reactor sites, thus relieving pressure to open a permanent repository at Yucca Mountain. Nevada also employs scientists who continue to review technical issues at the Yucca Mountain site; they have filed briefs and responses to items such as the draft EPA standards and the Environmental Impact Statement. This holds up the siting process because Energy often must take time to respond to the issues raised by the scientists.

Anti-Yucca opponents argue that politics have overtaken science. As Reid states on his Senate website, the 1987 NWPA amendment put undue political pressure on opening a repository at Yucca Mountain: "Since then, [Energy's] mission has shifted from objectively evaluating whether a site was suitable to isolate radioactive waste to justifying Yucca Mountain as a safe site for storing nuclear waste."

By singling out Yucca, the revised act pressured Energy to certify the site, and Congress wants to avoid revisiting the issue of site selection because the political costs are extremely high—no politician wants to allow a nuclear waste dump in his or her backyard. The nuclear industry is also eager to solve the nuclear waste problem, which it sees as an impediment to expanding the industry.

The original NWPA attempted to be fair in repository siting. It provided for two repositories to share the burden of nuclear waste among states and regions. Its provision to characterize three sites simultaneously ensured that no one site would bear the political pressure that Yucca Mountain now does. As a Utah newspaper editorial stated in 1981, "Neither Utah nor any other state can properly refuse to bear the nuclear waste burden once [the repository site] has been established to the best of human conditions. However, the honor of making such sacrifice for time without end must confer on

The Site of the Proposed Yucca Mountain Nuclear Storage Facility

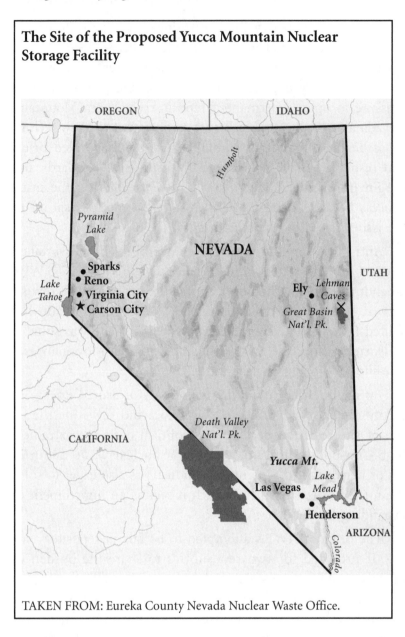

TAKEN FROM: Eureka County Nevada Nuclear Waste Office.

the luckless lamb the satisfaction of knowing firsthand that the duty couldn't have been just as well assigned elsewhere."[1] These comforts were lost with the passage of the 1987 amendments to the act.

Another political problem is funding. As of 2004, Energy had spent more than $8 billion studying Yucca Mountain. The money has come in part from the Nuclear Waste Fund, established by the NWPA, into which ratepayers pay one-tenth of one cent per kilowatt-hour of electricity used. By fiscal 2004, the fund had collected about $22.5 billion. Energy estimates that completion of Yucca Mountain would cost at least $60 billion, most of which would come from the fund, the difference made up by the Defense Department for disposing of its nuclear waste. The problems come in the allocation of funds for work on Yucca Mountain. When Congress sets its budget for the Energy Department, it has frequently not fully funded Energy's request for the Yucca Mountain project due to political influences. The money allocated for the project goes toward further characterization studies and facility construction; if the project were fully funded, the license application might proceed at a faster pace.

Formulating policy is always difficult when confronted with unknown variables. In the case of nuclear waste disposal, those unknown variables stretch over periods of geologic time. The proposed EPA standard requires that we understand repository behavior well enough 10,000 to 1 million years in the future to make accurate predictions of whether the radiation dose standard will be met. This is no small task.

The repository is essentially an "Earth system"—engineered features that will operate in a geologic environment. To understand how this system will behave over time requires predicting geologic processes and events. But geology is basically a retrodictive science, one that explains the past; it does not predict future events, such as volcanic eruptions or earthquakes, with accuracy. Using geology to make predictions is an inexact tool and will by definition produce highly uncertain results.

But predictions are useful for policy makers, who see them as a way to legitimize technical policies.[2] Moreover, focusing on technical predictions obscures other aspects of the debate, including economics, values, ethics, fairness, aesthetics, ideology, and local politics.[3] The state of Nevada, for example, makes its case against Yucca Mountain on a technical basis; there is little discussion of the issues of equity and regional politics, though these are central to Nevada's objections.

To understand the debate better, it is important to be familiar with some of the technical issues faced by Yucca Mountain and how the site is being evaluated. Energy and the NRC will decide whether Yucca Mountain meets the EPA dose standard by using a "probabilistic performance assessment." Performance assessments originate from analyses of nuclear reactors. Though these models may be appropriate for understanding the behavior of an entirely engineered system like a nuclear power plant over its lifetime of 40 to 60 years, it is problematic to apply them to a geologic repository over hundreds of thousands of years.

The complex performance assessment model for the Yucca Mountain site is made up of a number of submodels that simulate different aspects of the system. But the problem with modeling a geologic repository in this manner is that the models cannot be validated or verified. Earth systems are open systems, and as such there is no way to know all the input parameters or processes that might affect the system over time.[4] To use a phrase popularized by Defense Secretary Donald Rumsfeld in a different context, it is the "unknown unknowns—the ones we don't know we don't know"—that make validation of an Earth system model like the Yucca Mountain repository impossible. Despite this, Energy maintains it understands all the "features, events, and processes" that will affect repository behavior.

Furthermore, predictions of Earth systems have a reputation for inaccuracy and unreliability, often because they were based on bad assumptions.[5] Another shortcoming of performance assessment models is that different modelers working on the same project often arrive at different conclusions. (An International Atomic Energy Agency study on performance assessment models looked at six groups working on a model for the same subject; each group came up with very different results.)[6]

There are many technical hurdles involved with predicting whether Yucca will meet the EPA dose standard. Much of the scientific analysis of the site has concerned water. In large part this is because when spent nuclear fuel is exposed to moisture and air—the type of environment expected to exist in the Yucca Mountain repository—it becomes unstable and corrodes or "rusts," potentially releasing carcinogenic radionuclides. Conversely, spent fuel is stable in a wet and chemically reducing environment, such as exists below the water table; this is the type of repository environment that most other countries are pursuing for high-level waste storage.

If the Yucca Mountain repository can keep out all moisture, there will be little corrosion of the waste canisters and the spent fuel they contain. For years, Energy operated on the assumption that the water flow (the "percolation flux") from the surface to the water table at Yucca Mountain was very slow, about 4 millimeters per year or less. But in the mid-1990s, scientists from Los Alamos National Laboratory discovered unusually high levels of the isotope chlorine 36 in rocks at the repository level at Yucca Mountain. The high levels of chlorine 36 resulted from atmospheric testing of nuclear weapons over the Pacific Ocean from the 1940s to the 1960s. The presence of the bomb-pulse isotope, which drifted over land and precipitated down, suggests that water traveled very quickly—about 200–300 meters from the surface to the repository level in less than 50 years. This demonstrates the existence of "fast pathways"—likely faults and fractures—along which water can move rapidly through the rock. This discovery forced Energy to alter its conceptual model for water transport at Yucca Mountain. Just how much water moves along these fast pathways, and where they are located, is not yet understood.[7]

The percolation flux depends not only on the presence of fast pathways but also on the amount of precipitation, which is controlled by climate. To predict future precipitation, Energy considered climate at Yucca Mountain back to 400,000 years ago, encompassing periods that had more precipitation than today. What Energy's models do not include are predictions about global climate change that may occur over the next few hundred years. In 2100, carbon dioxide concentrations in the atmosphere may reach levels not experienced since 50 million years ago. Such an increase would likely be accompanied by major temperature increases and climate changes—the future climate at Yucca may well be much wetter than Energy is planning for.[8]

One of the most complex predictions concerns how the geochemical environment in the repository will evolve over time. Energy intends to use a "hot repository" design, keeping the temperature of the rock around the nuclear waste above the boiling point of water for hundreds of years, in order to keep moisture away from the waste canisters as long as possible. But over time, the heat and radiation from the nuclear waste will affect the rocks (by changing mineral compositions and creating fractures), the water in the rocks, the waste containers, and the waste itself. These interactions are basically impossible to predict.[9]

Another difficult prediction involves the tectonic processes that will affect the site over the next 1 million years. Yucca Mountain is located in a seismically and volcanically active area; volcanism poses the greatest threat to the ability of a repository at Yucca Mountain to contain radioactivity. It is possible that volcanic activity could result in the extrusion of magma, corrosive gases, and water into the repository tunnels. Less probable but more disastrous would be for an actual eruption to disturb the repository, spewing radioactivity into the atmosphere. But because there is a dearth of data on which to base predictions of future volcanic activity, the threat posed by volcanic activity is highly uncertain.

Partly in response to the slow progress on Yucca Mountain, the Bush administration is pushing a program, dubbed the Global Nuclear Energy Partnership (GNEP), which would revive the reprocessing of spent fuel. In congressional testimony on March 9, Energy Secretary Samuel Bodman described the project: "GNEP is a comprehensive strategy to enable an expansion of nuclear power in the U.S. and around the world, to promote nonproliferation goals, and to help resolve nuclear waste disposal issues."

Bodman also called Yucca Mountain "a complement to the GNEP strategy." But as a waste solution, GNEP has big problems: Some technologies the plan depends on, such as pyroprocessing and advanced nuclear power plants, will not be available for decades, if at all. Furthermore, existing GNEP reprocessing technologies will still create high-level nuclear waste as well as very large volumes of low-level and transuranic waste, which must, of course, be stored somewhere. The Bush plan simply defers to a later date the problem of dealing with high-level nuclear waste, when what is needed is immediate work toward an answer. Republican Cong. Joe Barton of Texas summed it up in March, directing this comment to Bodman at the budget hearing: "I am concerned, though, that the scope of this problem may be too broad and it may be premature. I would urge you not to allow the Global Nuclear Energy Partnership to divert focusing resources away from the near-term challenges that must be overcome to ensure the long-term viability of the industry, especially progress at Yucca Mountain."

Despite years of study and congressional mandates, it has yet to be determined whether Yucca Mountain is a suitable site. Answering this outstanding question is the imperative next step. What's needed now is more research—based on methods other than performance assessment modeling—to finally determine whether or not Yucca Mountain is a viable repository. Yucca should be evaluated in a comparative sense, as was originally planned in the Nuclear Waste Policy Act. It would be helpful to contrast Yucca with data on other well-studied sites worldwide; such locations could include New Mexico's Waste Isolation Pilot Plant (which stores transuranic waste from the nuclear weapons complex, and proposed repositories in Sweden, Finland, and France. Though it may simplify decision making for policy makers, performance assessment modeling is not a legitimate way to evaluate the site. For a complex Earth system like Yucca Mountain, the results of performance assessment modeling masquerade as quantitative analysis, whereas in reality they are riddled with subjectivity.

If Yucca Mountain doesn't measure up under a new comparative analysis, Congress will need to swallow hard and face the siting issue again. Instilling a sense of fairness in the legislation will help enormously, as will a look back at the original NWPA. Looking abroad can also serve as good guidance; France, Sweden, Finland, and Germany, all of which intend to open repositories, plan to involve citizens from the affected municipalities in the siting process. This consultative approach is in stark contrast to the U.S. "decide, announce, defend" policy.

Selecting a site for a national nuclear waste repository is one of the most difficult examples of policy making; there's a lot of pain per pound for the politicians involved. Despite problems encountered at Yucca Mountain, a geologic repository is by far the best solution to the nuclear waste problem. Even if a repository at Yucca is completed, politicians won't be able to avoid the siting issue forever. If Yucca were to open, it would be filled close to capacity with waste that has been temporarily stored around the country. Another repository would have to be opened to store future waste. The debate is not nearly close to being over—it's merely a harbinger of what is yet to come.

1. Quoted in E.W. Colglazier and R. B. Langum, "Policy Conflicts in the Process for Siting Nuclear Waste Repositories," *Annual Review of Energy*, 1988, vol. 13, pp. 317–357.

2. Daniel Sarewitz and Roger Pielke Jr., "Prediction in Science and Policy," *Technology in Society*, April 1999, vol. 21, pp. 121–133.

3. Charles Herrick and Daniel Sarewitz, "Ex Post Evaluation: A More Effective Role for Scientific Assessments in Environmental Policy," *Science, Technology, and Human Values*, Summer 2000, vol. 25, pp. 309–331.

4. For a discussion see Allison Macfarlane, "Uncertainty, Models, and the Way Forward in Nuclear Waste Disposal," in Allison Macfarlane and Rodney Ewing, eds., *Uncertainty Underground: Yucca Mountain and the Nation's High-Level Nuclear Waste* (Cambridge, Massachusetts: MIT Press, 2006). One of the first detailed discussion of this issue can be found in Naomi Oreskes, Kristin Shrader-Frechette, and Kenneth Belitz, "Verification, Validation, and Confirmation of Numerical Models in the Earth Sciences," *Science*, February 4, 1994, pp. 641–46.

5. For references, see Allison Macfarlane, "Uncertainty, Models, and the Way Forward in Nuclear Waste Disposal."

6. I. Linkov and D. Burmistov, "Model Uncertainty and Choices Made by Modelers: Lessons Learned from the International Atomic Energy Agency Model Intercomparisons," *Risk Analysis*, December 2003, vol. 23, pp. 1297–1308.

7. June Fabryka-Martin et al., "Water and Radionuclide Transport in the Unsaturated Zone," in *Uncertainty Underground*.

8. Mary-Lynn Musgrove and Daniel Schrag, "Climate Change at Yucca Mountain: Lessons from Earth History," in *Uncertainty Underground*.

9. See G. S. Bodvarsson, "Thermohydrologic Effects and Interactions," and William Murphy, "The Near Field at Yucca Mountain: Effects of Coupled Processes on Nuclear Waste Isolation," both in *Uncertainty Underground*.

| "The technology of radioactive waste 'disposal' . . . remains, obviously, experimental."

Safe Low-Level Radioactive Waste Storage Is Impossible

Judith Johnsrud

In the following viewpoint, Judith Johnsrud, director of the Environmental Coalition on Nuclear Power, recommends that environmental activists work toward promoting minimization of the production of all radioactive waste. All such waste, even low-level waste, presents a danger to humans and other species, she argues. Moreover, most of this waste remains dangerous for centuries; there is no way we can devise a technological solution to radioactive waste storage that will be safe under all possible future natural and social conditions, in Johnrud's view.

As you read, consider the following questions:

1. According to Johnsrud, why should the generation of radioactive waste be curtailed or ended?

2. Why should environmental activists *not* become advocates for particular "solutions" to specific waste storage problems, in the author's opinion?

Judith Johnsrud, "Low-level Nuclear Waste (LLNW) Management," Sierra Club, November 2006. www.sierraclub.org. Reproduced from sierraclub.org with permission of the Sierra Club.

3. According to Johnsrud, how many years have passed without the development of a successful program for dealing with radioactive waste?

Congress passed the Federal LLRW Policy Act in 1980, following the 1979 Three Mile Island accident, which created large amounts of unanticipated "low-level" radioactive waste (LLRW), and after objections were raised by the governors of the three states (South Carolina, Washington, Nevada) in which the nation had been dumping all of its commercial LLRW into commercial shallow land burial trenches. Several contamination events, plus increasing quantities of LLRW as more reactors came into operation, led them to assert that other states should share the burden. Three sites (in New York, Illinois, and Kentucky) had already been closed due in part to leakage. The Act was modified in 1985.

The law mandated that each state must "provide for" the disposal of all "low-level" wastes generated within its boundaries. It could construct disposal facilities to accommodate the LLRW produced within its jurisdiction, or arrange for shipment to a site in another state. To encourage development of disposal sites but also limit the total number, the Act also bypassed the Interstate Commerce Clause of the Constitution, allowing those states that formed compacts to exclude from a regional compact facility the "low-level" radioactive wastes (but not necessarily radioactive materials) generated outside the compact region. The law also required states to take title to LLRW; this provision was challenged by New York and overturned.

Members of Congress had been led to believe that most "low-level" wastes were generated by medical and research facilities and did not originate from commercial nuclear power plants. In fact, the opposite is true: in most states, more than 75% of the volume and more than 95% of the radioactivity of so-called "low-level" wastes are produced by nuclear reactors.

The term "low-level" has caused decision-makers, media, and the public to assume that LLRW consists of relatively harmless wastes: trash.

However, the term "low-level" does not mean "low hazard" to human health. All exposures to ionizing radiation, including naturally-occurring background radiation, carry risks to the recipient of somatic injury—e.g., leukemia, latent cancers, heart disease, and, it is now thought, immune system dysfunctions—as well as genetic damage, both physical and mental abnormalities. Moreover, there has been little consideration of the synergistic relationships of radiation and other environmental contaminants upon an individual recipient.

Although "Class A" wastes are composed mainly of low activity trash, some components may be biologically dangerous in minute quantities, and some remain hazardous for many thousands of years. The wastes deemed "Classes B and C" are higher in radioactive concentrations and tend to contain isotopes that have very long hazardous lives. Some LLRW may be declared "Greater Than Class C" in radioactive concentration and toxicity: these wastes are to be disposed of by the Department of Energy (DOE) as if they were high-level waste. The law categorizes essentially all nuclear wastes as "low-level," except for "spent" reactor fuel, some reprocessing waste, and whatever else the Nuclear Regulatory Commission (NRC) chooses to designate as "high-level" waste, plus certain byproduct materials, weapons-related wastes, and uranium mill tailings.

For states that choose not to join a compact, there is little legal precedent as to whether or not a non-compact state can exclude wastes generated beyond its borders. The Federal Act did not address the importation of radioactive materials that might subsequently be determined by a licensee to have no further economic value and hence be declared to be "waste." Nor was the LLRW Policy Act clear about the disposition of wastes in the hands of brokers, handlers, incinerator and treatment facilities: Were radioactive wastes from decontaminated materials and radioactive ash to be returned to the licensee and state of origin; or could they be considered commercially-generated wastes eligible for disposal within the state or compact in which the incineration or decontamination took place? Also unclear is eligibility for disposal of wastes imported from abroad: NRC promulgated import/export regulations only [recently]. . . .

Current Status of LLRW Management, Storage, and "Disposal"

Since passage of the LLRW Policy Act and its 1985 Amendments, no new LLRW disposal facility has been opened in the United States. Public opposition has repeatedly blocked LLRW siting in New York, Illinois, Nebraska, North Carolina, Vermont, Connecticut, Michigan, Colorado, Texas, Pennsylvania and other Host States. Even before temporary closure of Chem Nuclear's Barnwell, South Carolina, site in 1994 (as well as the U.S. Ecology site at Beatty, Nevada), the nuclear industry had responded to steeply rising disposal costs by minimizing volume of waste generated (but not activity), by storing to decay onsite, compaction, incineration, and decontamination treatment.

The NRC has tried since 1980 to resolve part of the LLRW problem by deregulating as much as one-third of Class A waste, variously called de minimis ("trivial," as in "de minimis non curat lex": "The law is not concerned with trivialities") or "Below Regulatory Concern" (BRC), and most recently termed "Incidental Radioactive Material." The agency intended to permit LLRW dumping in municipal solid waste landfills (as it already allows for some medical waste, sewage sludge, smoke detectors and other low activity wastes), more radioactive liquids into sewers, and especially the recycling of unmonitored, unlabeled "low-level" wastes into a wide array of consumer products and nuclear industry practices. The NRC's deregulation plan was thwarted by the 1992 Energy Policy Act, but is now being revived. The Energy Department has adopted "low-level" waste recycling.

Now that disposal costs at the reopened Barnwell burial site have risen to $350–$400 per cubic foot and are expected to go even higher, industry demands are increasing again for NRC deregulation. Even more important are the drive to discredit the linear non-threshold relationship of dose to response, which is the basis for radiation protection standards, and the effort to reduce governmental agencies' budgets by eliminating part or all of federal and state regulatory programs. The NRC's 1995 proposal to close down its LLRW division altogether is particularly significant, because NRC has preemptive power that overrides states' regulatory controls, and it conducts waste isolation research that the states can't afford. It regulates DOE, requires compatibility between federal and state programs, and licenses LLRW imports and exports.

One regulator now suggests a methodology for prioritizing agency expenditures for regulatory control of radiation exposures. It is based on dollars spent per life saved. An industry proponent offers a solution adapted from air pollution credits trading: an "Open Market Trading Rule" for radiation doses; let the affected community decide for its population if it wants to pay for reducing risks of fatal cancers and other adverse health and genetic effects from the various sources of ionizing radiation and other contaminants encountered by individuals in their environment, life styles, or medical treatment.

As waste volumes (but not radioactivity) decreased, the "crisis" need for centralized disposal sites in the ten compacts and eight unaffiliated states also has diminished. Better housekeeping by generators, NRC's waffling about the duration of onsite storage, possible LLRW export, and emergence of the privately-owned Envirocare site in Utah for surface disposal of low activity wastes have also reduced the urgency. The NRC and industry now estimate that three or four sites will probably be sufficient. If a few disposal sites are finally opened, Congress may be asked to declare those the national sites.

However, it is not known how much more, or less, "low-level" waste will be added from future decommissioning of aged nuclear power, research and naval reactors; possibly from weapons-related DOE or other military sources; and from the remediation of the more than 45,000 sites radioactively contaminated or potentially contaminated sites that have been identified by the Environmental Protection Agency. Increasingly, as budgets tighten and costs of waste isolation rise, regulators are considering decommissioning criteria based on "How dirty is clean enough?" They would permit licensees to leave behind a still-contaminated site for "restricted" "brownfield" use. In 1994, NRC staff proposed for "a few tens" of heavily contaminated sites, onsite stabilization and disposal "despite the failure to meet the 100 millirem per year [dose] cap" [to the average member of the critical group of those expected to be exposed]. The final decommissioning criteria are to be promulgated in the near future.

The technology of radioactive waste "disposal" for even just the 300–500 years required for Classes B and C, remains, obviously, experimental. Chem Nuclear has continued to use shallow trench burial at Barnwell, despite commitments to build above grade, mounded, retrievable, monitored storage vaults. French vaults, containing long-lived wastes, have been in service fewer than ten years. Industry consultants are already arguing that occupational doses will be lower with shallow land burial than above grade vaults due to less handling. But stability of waste forms, for example, remains in controversy. A June 1996 report on microbial degradation of cement issued by NRC states:

Testing conducted with the developed biodegradation test has convincingly demonstrated that cement-solidified LLW

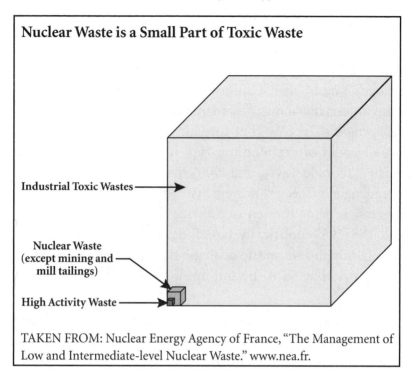

Nuclear Waste is a Small Part of Toxic Waste

Industrial Toxic Wastes

Nuclear Waste (except mining and mill tailings)

High Activity Waste

TAKEN FROM: Nuclear Energy Agency of France, "The Management of Low and Intermediate-level Nuclear Waste." www.nea.fr.

waste forms can be attacked and degraded by the action of ubiquitous microorganisms that are present at LLW disposal sites. It was shown that during the degradation process, large percentages of those elements composing the cement matrix of waste forms were removed. In addition, it was conclusively shown that the ability of cement-based waste forms to retain or retard the loss of encapsulated radionuclides was compromised due to the action of microorganisms. (NUREG/CR-6341; INEL-95/02i5)

Moreover, some researchers are joining environmentalists in recognizing a simple fact of basic physics: we don't "dispose of" anything; we can only change the locations or forms of matter. This realization is sparking the demand for an independent reconsideration of all of the nation's radioactive waste programs—and even legitimizes the call by Sierra Club and others for curtailment of waste production.

How Compacts and States Are Doing

At mid-year 1996, both Chem Nuclear and U.S. Ecology admit to financial problems, and LLRW siting is still in disarray. U.S. Ecology's Ward Valley, California, desert site near Needles remains uncertain; industry and regulators believe it is key to their program's success, but Interior Department's land transfer and issues of a plutonium cap, radiation pathways to the nearby Colorado River, and concerns for the desert tortoise, an endangered species, prevent its completion and operation. Congress is being pressed to override these concerns, as it did with the 1987 politically-based designation of the Yucca Mountain high-level waste geologic repository site in Nevada, which appears also to be failing to meet health and safety standards.

At the closed U.S. Ecology Beatty, Nevada, site, tritium contamination has been found. An upcoming report on the six older LLRW dump sites omits all data after 1994; hence, no mention of Beatty leakage. U.S. Ecology's proposed Boyd County, Nebraska, site has also been delayed by admission that wetlands had been ignored in site characterization. Its older Richland, Washington, site now accepts LLRW from only the Northwest and Rocky Mountain Compacts; earlier, U.S. Ecology had also been hit with some heavy expenses by the state . . . [and] the company was reportedly near a declaration of Chapter 11 bankruptcy.

Similarly, Chem Nuclear (CNSI) has a disappointing record at the reopened Barnwell; its North Carolina site is stalled by lack of funds. With more than $90 million reportedly spent, some $26 million more are needed for the site approval process. On July 10, North Carolina terminated its contract with Chem Nuclear to construct a LLRW facility for the Southeast Compact at the Wake County site, according to a July 18th report and the June/July 1996 LLW Forum Notes (from DOE funded Afton Associates). Almost all CNSI workers were dismissed, except for a skeletal crew that was assigned to remove equipment and structures from the site.

Illinois, after failure of Chem Nuclear's Martinsville volunteer site, is still revising its siting criteria and must start over. In Pennsylvania, a new Republican Administration has abandoned the strict technical siting process to identify three "of the best" locations, as is required by law, in favor of Chem Nuclear's volunteer "Community Partnering Plan." Environmentalists are blamed for encouraging counties and municipalities to adopt protective ordinances and to place land in Agricultural Security Areas. State regulators threaten to "abbreviate" the remaining mandatory technical review, or to change the law. Connecticut, New York, and New Jersey volunteer processes have failed thus far to produce a LLRW site.

The contested Texas site at Sierra Blanca awaits a licensing decision; the Texas Compact with Maine and Vermont is before Congress but not of this date approved. Michigan has adopted onsite storage for the foreseeable future. Massachusetts has repeatedly delayed siting action. No one wants to be first.

Beyond Backyards

1. We all surely agree that the radioactive wastes we (all) have allowed to be produced for half a century pose a significant biological hazard to humans, to many other life forms and ecosystems; and that they must be isolated from the biosphere for the full duration of their hazardous lives. Yet, given the situation of ongoing, open-ended waste production and the many political and economic uncertainties of waste generation and regulation, quite rightly no one wants to offer his or her own backyard as the site for an endless, perhaps increasing, burden of long-lived nuclear wastes. A major ethical responsibility for us all is, after all, to the future as well as present survival and well-being of our species, our

descendants, and the other inhabitants of the planet, whether we depend on them or not. This is also our national policy, enunciated in the National Environmental Policy Act of 1969.

2. Those of us who have warned of problem for years—decades—are now told that it is we who are responsible for providing "solutions" that will allow the nuclear industry to continue to produce ever more radioactive waste. And because we want to seem reasonable, positive, or constructive, there is a great temptation to recommend relocating the problem: Send it to a desert wasteland; keep it onsite where people must have wanted reactors and where they supposedly benefit from the electricity; or airlift it to Dagestan or West Africa or Mexico where folks need the money: derive some economic benefit from it by recycling into consumer products (just a few more millirems from each one); trade off one risk for another. These are surely temptations to be resisted for all the reasons a good environmentalist understands.

3. But then what can we recommend?

- First, by working backward from the impossibility of assuring safe permanent isolation for the full period of nuclear waste toxicity to the cause of the problem (i.e., continued production of the waste), we'd have the opportunity to make the best reasonable, commonsensical case for a national policy of curtailing, with the intent of ending, the generation of most radioactive wastes. In the opinion of some, this should happen immediately, so that we will be able to assess the quantities, composition, and toxicities of what we must prepare to deal with.

- Second, we can help decision-makers to comprehend the disconnect between our present ability to assure full hazardous life sequestration and the realities of the future world about which we can only make our best guess about economies, political structures, scientific and cultural capabilities, the press of expanding populations on diminishing natural resources, climatic, tectonic, or other physical changes in the biosphere, possible improvement or decline of waste management—a total system (or geographic) analysis. This, in turn, leads us to require the greatest prudence and conservatism, given the limits of prediction and the experimental nature of the endeavor.

- Third, we can recommend that the focus of waste management be shifted away from the notion and technologies of permanent "disposal" to the real reason for the need to prevent radioactive materials and wastes from entering the environment: namely, the hazard posed to health, safety, genetic integrity, and the environment by exposures to ionizing radiation. Although at present no one can be sure of "the best" means or technologies or locations to assure minimization of the biologic damage from this unstable form of matter, we can urge that the prevention of exposures and reduction—minimization—of the biologically damaging consequences are the real societal goal.

- Fourth, we need to explain to decision-makers that some sites currently in use for storage of LLRW are particularly ill-suited for that purpose (i.e., due to potential seismicity, flooding, etc.), and thus do not meet the crucial test of "best means to assure minimization of biologic damage." If, by contrast, we become advocates for any one of the bad "solutions" (which can be expected to fail to contain the waste), we've become advocates of imposing on others the risk of LLRW isolation failure that we find unacceptable in our own backyards. Such advocacy also gives waste generators and regulators an excuse of claiming that they are only responding to what "the public" wants.

- Fifth, we can suggest ways to achieve that goal. For instance, the regulators can, and should, set environmental pollutant standards that take into account the multiple and cumulative exposures to the many contaminants that affect an individual. Governmental agencies should expand our understanding of the synergistic relationships among the variety of pollutants to which individuals are subjected and take these interrelated factors into account in the setting of "routine permissible" dose limits. And the burden of proof of safety must be assigned to the waste generators, not to those who are damaged.

We can empower communities to monitor contaminants and exercise greater control over their generation and release into air, water, and soil. Even though some waste would be produced in the process, it might be feasible to encourage waste generators to seek less damaging ways of earning a profit, by beginning to set annually permissible release limits for environmental contaminants that are increasingly restrictive, moving always toward the zero release goal, requiring improved pollution control year by year, and offering companies opportunity to adjust their investments and production activities to more benign ends. We can even suggest that there are more beneficial ways of perceiving reality than our culture's near-religious faith in technology's ability to solve all environmental problems permits us to use.

Idealistic dreaming? No. Impossible to achieve? Only in the short term. For, as dedicated environmentalists, we surely understand that the ultimate survivability of life on earth does depend on our willingness to move our political, economic, and social structures into conditions of greater compatibility with the earth. In the face of fifty years of failure to dispose of radioactive wastes, perhaps there are even some decision-makers who are now willing and able to listen to what we have to tell them.

"Scientists and engineers have the knowledge and ability to design and build facilities to contain [low-level radioactive] waste safely for the required period and longer."

Low-Level Nuclear Waste Can Be Disposed of Safely

Nuclear Energy Institute

The Nuclear Energy Institute is an organization that advises the public and policy makers on matters affecting the nuclear energy and nuclear technology industries. It advocates the safe use of nuclear power to meet America's energy needs. In the following viewpoint, the institute describes methods of handling and disposing of low-level radioactive waste and makes the case that adequate regulations and tecnnologies exist for the safe handling and disposal of such waste.

As you read, consider the following questions:

1. According to the institute, what are some of the activities that generate low-level nuclear waste?

2. How many types of low-level waste are defined by Nuclear Regulatory Commission regulations, according to the author?

Nuclear Energy Institute, "Fact Sheet: Disposal of Low-Level Radioactive Waste," August 2004. www.nei.org. Reprinted courtesy of NEI.

3. What three types of waste *disposal* does the institute name?

Many socially beneficial activities use radioactive materials, producing low-level radioactive waste as an unavoidable by-product. These activities include electricity generation, diagnosis of illness without exploratory surgery, treatment of diseases like cancer, medical research, testing of new pharmaceuticals, nondestructive testing of airplanes and bridges, hardening of materials like hardwood floors, the breeding of new varieties of seed with higher crop yields, eradication of insect pests, food preservation, ionization-type smoke detectors and dozens of other purposes.

America's nuclear power plants are the nation's second-largest source of electricity, generating about one-fifth of U.S. electricity—without producing greenhouse gases or contributing to air pollution.

As a byproduct of their operation, these plants generate more than half the volume, and most of the radioactivity, of the nation's low-level waste.

The remaining low-level waste is generated by several thousand other industrial facilities and institutions that use radioactive materials—medical research laboratories, hospitals, clinics, pharmaceutical companies, government and industrial research and development facilities, universities and manufacturing facilities.

What Is Low-Level Waste?

Low-level waste includes items such as gloves and other personal protective clothing, glass and plastic laboratory supplies, machine parts and tools, filters, wiping rags, and medical syringes that have come in contact with radioactive materials.

Low-level waste from nuclear power plants typically includes water purification filters and resins, tools, protective clothing, plant hardware, and wastes from reactor cooling water cleanup systems.

Low-level waste is solid material. It generally has levels of radioactivity that fade to levels in the general environment in less than 500 years. About 95 percent fades to these background levels within 100 years or less. Low-level waste does not include used fuel from nuclear power plants or high-level waste from nuclear weapons reactors.

NRC [Nuclear Regulatory Commission] regulations separate low-level waste into three classes: A, B and C. The classification of the waste depends on the concentration, half-life and types of the various radionuclides it contains. The NRC sets requirements for packaging and disposal of each class of waste.

Class A low-level waste contains radionuclides with the lowest concentrations and the shortest half-lives. About 95 percent of all low-level waste is categorized as Class A.

Classes B and C contain greater concentrations of radionuclides with longer half-lives. They must meet stricter disposal requirements than Class A waste.

Low-level waste that exceeds the requirements for Class C waste—known as Greater than Class C waste—is, under federal law, the responsibility of the U.S. Department of Energy [DOE]. This material is less than 1 percent of all low-level waste. . . .

Successful Waste-Minimization Procedures

Since 1980, because the cost of low-level waste disposal has been rising and access to disposal sites has been in jeopardy, waste generators—industry, government, utility, academic and medical—have dramatically reduced the volume of low-level waste sent to commercial disposal sites. In 1980, more than 3.7 million cubic feet of low-level waste were disposed of commercially. In 2001, the volume of "traditional" low-level waste (higher radioactivity levels, low-volume) declined to 140,147 cubic feet, a reduction of 96 percent. This reduction occurred even though the number of nuclear power plants increased by more than 50 percent during the same period.

In addition to the traditional low-level waste disposed of in 2001, more than 1.2 million cubic feet of low-activity, high-volume, low-level waste was disposed of at the Envirocare facility in Utah from the decommissioning of nuclear facilities and site cleanup activities.

Volumes of traditional waste have dropped because generators have worked hard to produce less of it. Management practices have improved—for example, segregating radioactive and nonradioactive materials. In some cases, nonradioactive materials have been substituted for radioactive materials.

Procedures for Waste Handling

Generators also have been successful in reducing the volume of low-level waste already generated. Techniques include compaction, incineration, decontamination and storage while it decays.

On-site decay Many radionuclides in low-level waste decay to safe levels within a relatively short time. When wastes are safely stored at their generation sites for a few days to a few years (depending on half-life and available storage space), the radioactivity may be reduced to safe background levels. On-site storage of low-level waste is regulated by the NRC and/or the agreement states.

Compaction Compactors can achieve as much as a 90 percent volume reduction, depending on the size of the compactor and the type of waste. Today's "supercompactors," which can exert a force of 1,000 tons or more, can crush dry waste to a small fraction of its original volume.

Decontamination By decontaminating large pieces of equipment, tools, metal, glassware and clothing, low-level waste generators are able to reuse or recycle them.

Incineration Combustible dry wastes can be incinerated with as much as a 99 percent reduction in volume. The NRC or agreement state strictly regulates how much radiation can be

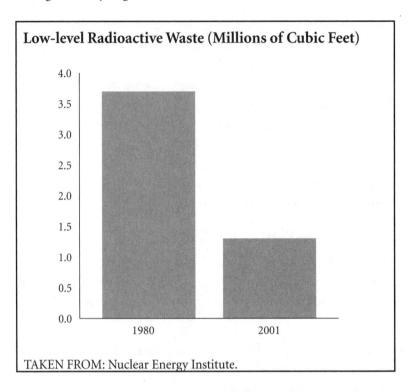

Low-level Radioactive Waste (Millions of Cubic Feet)

TAKEN FROM: Nuclear Energy Institute.

released to the environment through incineration. Incinerators are equipped with filters and other technologies that make sure these limits are strictly adhered to and the public protected. Virtually all the radioactivity in incinerated materials remains in the ash, which is disposed of in licensed low-level waste disposal facilities.

Safely Transporting Low-Level Waste

There are over 300 million U.S. shipments of hazardous materials annually, about three million of which involve radioactive materials such as radiopharmaceuticals or radioactive compounds for medical research. Only a small fraction of these shipments contain low-level waste.

From 1973 to 1995, there have been only four low-level waste transportation accidents that led to the release of radioactive material—for example, a package opening. In all cases, the releases were small and the released materials were quickly repackaged. No injuries or deaths have ever been caused by a release from low-level radioactive waste in a transportation accident.

The NRC requires that radioactive materials be packaged for shipment to protect the public in case of an accident. The kind of packaging required depends on the amounts and types of radioactive elements in the waste.

Stringent standards for containers Low-level waste is shipped in containers designed to meet stringent NRC and DOT [Department of Transportation] standards. Most low-level waste contains low enough levels of radioactivity to be shipped in strong, tight containers or DOT Type A containers. (Type A and B shipping containers bear no relation to NRC Class A, B and C waste forms.) Type A containers must be able to withstand ordinary transportation conditions.

Wastes containing higher levels of radioactivity are shipped in Type B containers, which must be able to withstand accident conditions. Tests must demonstrate that shipment casks can survive a 30-foot fall onto a flat, unyielding surface; a 40-inch drop onto a six-inch steel spike; a 30-minute exposure to a fire of 1,475 degrees Fahrenheit; and submersion in 50 feet of water for eight hours.

Accident response arrangements Besides these precautions, computer programs can be used to select the safest transportation routes. Carriers receive emergency response training that includes written procedures. In the unlikely event of a serious accident, the Federal Emergency Management Agency has instituted the Federal Radiological Emergency Response Plan in concert with 11 other government agencies. Local police, firefighters and state radiological protection teams also are trained to respond in an emergency.

Nuclear plant operators have signed a nationwide agreement providing that, in case of an accident involving radioactive materials, the closest company will provide equipment and technical assistance to the emergency response team, regardless of who shipped the radioactive material. . . .

Government Regulation of Radioactive Waste

The NRC and the states govern the siting, operation and closure of all low-level waste disposal facilities. The purpose of low-level waste disposal is to isolate the waste from people and the environment. The NRC has set forth requirements to protect people from releases from the site, prevent inadvertent intrusion into the waste, protect workers during operation and ensure the stability of the site after closure.

The NRC regulations ensure that Class A low-level waste is contained for up to 100 years, Class B waste for 300 years and Class C waste for up to 500 years. Scientists and engineers have the knowledge and ability to design and build facilities to contain the waste safely for the required period and longer.

The NRC has set forth technical requirements for low-level waste disposal sites. They require, among other things, that natural resources in the area, such as wildlife preserves, be avoided, the site must be sufficiently isolated from groundwater and/or surface water; and the site must not be in an area of geological activity (such as volcanoes or earthquakes).

Types of Waste Disposal

Regardless of design, all low-level waste disposal sites use a series of natural and engineered barriers to prevent radioactivity from reaching the biosphere. There are several designs for building disposal facilities.

Shallow land burial This method was used by all U.S. low-level waste disposal facilities until 1995. Waste containers are placed in long, lined trenches 25 or more feet deep. The trenches are covered with a clay cap or other low-permeability cover, gravel drainage layers and a topsoil layer. They then are contoured and replanted with vegetation for drainage and erosion control. In addition, an intrusion barrier, such as a thick concrete slab, is added to Class C waste trenches. The sites are carefully monitored to ensure performance in compliance with the regulations. Facilities are sited in an area away from surface water and where travel of any groundwater is slow. The sites are monitored to ensure that there is no leakage.

Modular concrete canister disposal This method consists of individual-waste containers placed within concrete canisters, which are then disposed of in shallow land sites. The array of canisters has an earthen cover. This additional engineered barrier system has been used at the Barnwell, S.C., facility since 1995 and has been proposed for the Class B and C disposal at Envirocare.

Belowground vault This type of disposal uses a sealed structure built of masonry blocks, fabricated metal, concrete or other materials that provide a barrier to prevent waste migration. It has a drainage channel, a clay top layer and a concrete roof to keep water out, a porous backfill, and a drainage pad for the concrete vault. This design has been used successfully at a DOE disposal site.

Aboveground vault or engineered berm This is a reinforced-concrete building that provides isolation on the earth's surface. Its walls and roof are two to three feet thick, and it has a sloping roof to aid water runoff. Some Canadian utilities use similar aboveground vaults for storing low-level waste for later disposal. For low-activity radioactive waste, aboveground engineered berms provide the same isolation as shallow land burial. Envirocare of Utah uses aboveground engineered berms.

Earth-mounded concrete bunkers In these facilities, wastes are placed in belowground, concrete monoliths, and less radioactive waste is placed on top of the monoliths to create the mounds. The bunker is equipped with a drainage system and covered with impermeable clay and topsoil, giving the facility a rounded shape. These have been successfully used in France for at least two decades.

Generally, all costs of site development, operation and long-term monitoring are paid by the generators of the waste.

No Threat to Health or Safety

The radioactive particles in low-level waste emit the same types of radiation that everyone receives from nature. Most low-level waste fades away to natural background levels of radioactivity in months or years. Virtually all of it diminishes to natural levels in less than 500 years.

The NRC requires that releases of radiation at a disposal site not exceed an annual dose to any member of the public of 25 millirem to the whole body, 75 millirem to the thyroid, or 25 millirem to any other organ. In comparison, the average American is exposed to about 360 millirem of radiation annually—mostly from natural sources (such as radon) and medical sources (such as X-rays). In practice, public exposures from low-level waste facilities are far lower than the NRC limits.

> "Reports of adverse health effects linked to the use of sludge as fertilizer have mounted."

Sewage Sludge Recycling Poses a Threat to Human Health

Caroline Snyder

Caroline Snyder taught environmental studies at the Rochester Institute of Technology for twenty years. In the following viewpoint she describes the potential dangers of treating agricultural and commercial land with sewage sludge. According to Snyder, sludge—the substance that remains after treating wastewater— contains a variety of biological and chemical hazards to human health. Snyder maintains that both state and federal governments have covered up the dangers of this waste product.

As you read, consider the following questions:

1. According to the National Academy of Sciences, as cited by Snyder, why is it impossible to assess the risks of using sewage sludge for farming?
2. Why was the original 503 Rule—the federal regulations governing sludge disposal—weakened, according to the author?

Caroline Snyder, "The Dirty Work of Promoting 'Recycling' of America's Sewage Sludge," *International Journal of Occupational and Environmental Health*, vol. 11, October-December 2005, pp. 415–27. © 2005 Abel Publicatio Services, Inc. Reproduced by permission of the publisher and author.

3. According to Snyder, how many deaths are linked with the land application of sewage sludge?

The United States Federal Clean Water Act defines municipal sewage sludge as a pollutant. Typical sludges from industrialized urban centers contain tens of thousands of contaminants, from industry, institutions, businesses, landfills, and households, that discharge into sewers. Wastewater treatment plants are designed to remove these pathogens, metals, and chemical compounds—many of which are toxic and persistent—from wastewater. Almost all the removed material, by necessity, concentrates in the resulting sludge. Every month, every industry in the country is permitted to discharge up to 33 pounds of hazardous waste into sewers without reporting.

Sewage Used As Fertilizer

Despite the fact that sewage sludge is a contaminated waste product, it is being commonly treated and used as a fertilizer, without informing the recipients about the complete contents of the sludge. In 2002, a National Academy of Sciences (NAS) panel warned that treated sewage sludge is such a complex and unpredictable mix of biological and chemical wastes that its risks, when used for farming, cannot be reliably assessed. Therefore, the panel concluded, standard strategies to manage the risks of land application do not protect public health.

Even though the effects of treated sludge are unpredictable, complex, and potentially harmful, the United States Environmental Protection Agency (EPA) has failed to appropriately manage its disposal. Instead, upper-level managers in the agency's Office of Water (OW) and Office of Research and Development (ORD) abandoned their agency's mission by yielding to industry pressure to promote and defend the risky practice of using a contaminated waste product as a fertilizer.

Reports of adverse health effects linked to the use of sludge as fertilizer have mounted, especially in the last ten years.

Over the same time, EPA forged a powerful alliance with municipalities that needed an inexpensive method of sludge disposal and sludge-management companies that profit from this practice. The alliance's primary purpose was to control the flow of scientific information, manipulate public opinion, and cover up problems, in order to convince an increasingly skeptical public that sludge farming is safe and beneficial. The alliance ignored or concealed reported health problems, threatened opponents with litigation, distributed misleading information to the media, legislators, and the public, and above all, attempted to silence critics.

Since 1996, EPA's efforts to silence opponents have been the subject of Labor Department investigations, congressional hearings, Inspector General audits, and lawsuits filed by farmers and residents. This article draws on these proceedings and other information by explaining how EPA uses industry-friendly scientists and corporate influence to defend an unprotective policy. It's a carrot-and-stick approach. Supportive scientists receive federal grants, while economic threats are used to silence unsupportive scientists, private citizens, and local governments.

The Government Promotes Sludge Use

Since its inception, EPA has been promoting sludge use for farming. In the late 1970s, the first land application regulations were formulated by managers and scientists in EPA's Office of Water (OW): Henry Longest II, John Walker, and Alan Hais. As Deputy Assistant Administrator of OW, Longest was one of the people responsible for administering the funds for EPA's multi-billion-dollar Construction Grants Program, the United States' largest public works project ever. The purpose of the project was to build wastewater treatment plants, as mandated by the Clean Water Act.

The rapid proliferation of new wastewater treatment plants produced vast quantities of sludge. And because industrial wastes that used to be dumped into rivers were now discharged into sewage systems, the sludge became much more hazardous, often qualifying as hazardous waste. At the time, the Resource Conservation and Recovery Act (RCRA) was being developed to regulate hazardous waste. During this inflationary period, municipalities demanded an inexpensive way of disposing of their sludge. President [Jimmy] Carter's appointee to OW, Thomas Jorling, insisted that sludge not be regulated under RCRA, and the Act was weakened. The watered-down Act allowed not only sludge but also industrial wastes to be legally used as fertilizer. . . .

A significant amount of the country's hazardous waste from industries and other institutions is in the form of wastewater. Under the domestic sewage exclusion, industries are permitted to discharge hazardous wastewater into sewer lines to mix with domestic sewage entering publicly owned treatment plants. The assumption that this wastewater has been adequately pretreated by the sources, so that sewage sludge contains only "low levels of toxic substances," [according to the NAS] has been widely questioned.

There were early warning voices within the agency that sludge and industrial waste used as fertilizer would lead to serious problems down the road. William Sanjour, chief of EPA's Office of Solid Waste Management Programs Technology Branch, warned repeatedly that Mr. Jorling's order to reduce the scope of RCRA so that sewage sludge and other industrial waste could be land applied "was illegal and inconsistent with the agency's congressional mandate to protect human health and the environment." Sanjour's warnings, however, went unheeded, and EPA removed him from his position.

Covering Up Environmental Hazards

The campaign to promote "beneficial use" of sewage sludge continued in the 1980s. It was becoming "a murky tangle of corporate and government bureaucracies, conflicts of interest,

© CartoonStock.com

and cover up of massive hazards to the environment and public health." In 1981, EPA published a document describing the various persuasion techniques that could be used to induce the public to accept land application. Preferred application sites were rural low-income neighborhoods where cash-strapped farmers were told municipal sewage sludge was superior to manure and commercial fertilizer, would dramatically increase yields, and, best of all, was free. EPA and wastewater treatment plants did not inform rural residents about the potential hazards that might occur from using this material.

The only thing missing at EPA was a body of scientific evidence that explained why chemical pollutants, considered toxic and regulated elsewhere, are somehow beneficial when present at the same or higher levels in processed sewage sludge. In 1987 Congress reaffirmed its 1977 directive that EPA develop "a comprehensive framework to regulate the disposal and utilization of sludge." The fact that EPA developed these regulations post hoc [after this] to justify an existing policy was problematic. Would the regulations be truly science-based and protective, or would they merely rationalize an existing policy?

The sludge-disposal problem became more urgent in 1988, especially in the Northeast. Sludge generated in coastal cities was being dumped into the ocean. This impacted marine organisms and damaged beaches. Outraged environmentalists succeeded in having Congress pass legislation prohibiting ocean dumping. Environmental groups unwisely agreed to sign a consent decree supporting land application if, in return, ocean dumping would stop.

From 1989 to 1992, land application was governed by a stringent interim rule, the 1989 proposal. In the absence of good science, this first version of the 503 rule included strict precautionary metal standards as well as standards for 12 toxic organic chemicals. The interim rule met with strong opposition from municipalities and sludge-management companies. Sludge generated in many large urban centers could not meet these strict standards. In addition, the extra testing requirements for toxic organics would be time-consuming and expensive. Cities that had depended on cheap ocean dumping insisted that disposal of sewage sludge should remain convenient and inexpensive. Also, hauling sludge from cities to nearby farms was becoming a growing and lucrative business. Robert O'Dette, representing the sewerage industry, warned in 1990 that if the interim rules were adopted, beneficial reuse of sludge would end.

Thus, pressure from municipalities and the sewerage industry ensured that the final rule, the 503 rule, would be so lenient that virtually all municipal sewage sludge could legally be land applied. . . .

Sludge Use Faces Scientific Criticism

In March 1997, the prestigious Cornell Waste Management Institute (CWMI) released a working paper "The Case for

Caution," which was revised in 1999 and published in a peer-reviewed journal under a different title. This was the first comprehensive science-based critique of the 503 rule. In their opening sentence the authors boldly state: "Current US federal regulations governing the land application of sewage sludges do not appear adequately protective of human health, agricultural productivity, or ecological health." Between April and December 1997, New York State regulators worked closely with Alan Rubin [an EPA scientist], John Walker, EPA's Assistant Administrator, and Rufus Chaney, of USDA [United States Department of Agriculture], on a response to the Cornell paper. Copies of their correspondence were sent to the President of Cornell University. On July 24, 1997, EPA's Assistant Administrator wrote to the Deputy Secretary of USDA: "I am quite concerned about the Cornell paper. We believe the publication being proposed by Cornell . . . will have a negative impact on the use of biosolids." Subsequently the nation's leading sludge-management company paid a group of sludge-friendly scientists to attack the paper. Cornell scientists, however, have not wavered in their critique of the 503 sludge rule.

At the same time, David Lewis, one of EPA's internationally known senior research scientists, began investigating reported cases of illnesses and deaths among sludge-exposed individuals and started to form a theory of why some of them were suffering serious health problems. Lewis presented his findings at various scientific meetings and began submitting the work to EPA managers for clearance as a series of research articles and commentaries in peer-reviewed scientific and technical journals. EPA managers in Washington, DC, and at Research Triangle Park, NC, responded by ending all of his research funding in 1998 and instructing his local supervisors in Georgia not to let him collaborate with other EPA scientists or let him have access to agency resources. He raised enough, including $80,000 of his own personal funds, to continue his sludge research until 2004. . . .

Illnesses, Deaths, and Denials

Meanwhile, hundreds of rural neighbors living or working adjacent to sludged fields reported unbearable quality-of-life conditions as the stench from this chemical and biologically active waste material forced them to retreat inside their homes. Many reported serious adverse health effects after being exposed to sludge. These included nausea, vomiting, burning eyes, burning throats, congestion, various infections, and serious respiratory problems. Others, including infants, had to be rushed to hospitals because they had trouble breathing. The three deaths linked to land application were those of Shayne Connor, from Greenland, NH, Daniel Pennock, from Robesonia, PA, and Tony Behun, from Osceola Mills, PA. While the parents of Shayne, Daniel, and Tony were mourning their sons' deaths, WEF [Water Environment Federation] distributed EPA-funded "fact sheets" with EPA assurances that there were "no documented cases of illnesses" and "no public health concerns from the use of biosolids whatsoever."

Tony Behun's death intensified the public concern over sludge application in Pennsylvania. For land application to continue under the current policies, it was essential for the Pennsylvania Department of Environmental Protection (PA DEP) to deny that sludge might have caused the death of a Pennsylvania child. Len Martin compiled a chronological and detailed account of how, for almost two years, the PA DEP went to extraordinary lengths to hide the circumstances of the child's death. In October 1994, 11-year-old Tony had ridden his dirt bike through sludge that had been applied to a reclaimed mining site. The child developed headache, sore throat, furuncles on one leg and arm, difficulty breathing, and a high fever. On October 21, a week after he had been exposed to sludge, Tony died of staphylococcal septicemia [bacterial infection of the blood].

In 1999, Tony's mother, who had heard that sludge was causing health problems in other parts of the country, sought answers from the state about her son's mysterious death. The PA DEP repeatedly and publicly denied that there was any connection between sludge exposure and her son's death. According to public statements made by the agency and the company that had spread the sludge, Tony's death resulted from a bacterial infection caused by a bee sting, and sewage sludge had not been applied on the mining site. In May 2000, PA DEP secretary, James Seif, drafted a report claiming that both the National Institute of Occupational Safety and Health (NIOSH) and the state health department had investigated the case thoroughly and ruled out sludge as the cause or contributing factor of Tony's death. Every one of the above-cited claims proved to be false. The DEP was forced to retract the fabricated bee-sting story; truck weigh slips indicated that about 5,600 wet tons of sludge had been spread on the site next to the child's home; and on August 7, 2000, the PA Department of Health sent a letter to State Representative Camille George confirming that the department "in fact, did not conduct an investigation into Tony Behun's death." NIOSH also stated that it had no involvement [in the case] because "our agency only investigates workers' health complaints."

Subsequent public testimony by EPA's Robert Bastian illustrates how EPA and the state agencies responsible for land-application policies work together to misrepresent facts to cover up incidents. On March 13, 2001, Bastian presented Seif's report to the NAS panel that was investigating information about alleged health incidents linked to sludge and assured the panel that "the findings of [PA] state and local health officials have indicated that the Pennsylvania death was not attributable to biosolids." . . .

EPA claims it no longer promotes land application of treated sewage sludge. Yet there is no indication that the agency has divorced itself from the industry it is supposed to regulate. . . . EPA and its alliance partners continue to assure increasingly skeptical audiences that land-applied sludge is "an extremely safe material."

> *"There [has] never been a demonstrated instance of human or animal disease resulting from biosolids that are recycled in accordance with regulations."*

Land Application of Biosolids Is Safe

Sam Hadeed

In this viewpoint, Sam Hadeed offers scientific evidence that biosolids, solid or semisolid material obtained from treated wastewater, do not threaten health either directly or indirectly. Hadeed is technical communications director for the National Biosolids Partnership, an organization that serves the biosolids and sewerage industry. He stresses the need to continue scientific data gathering at biosolids application sites and to keep communities informed of findings. He believes that information collection and sharing will allay the public's fears about using sludge as fertilizer.

As you read, consider the following questions:

1. According to Hadeed, what are the most common concerns of the public regarding application of biosolids?

2. What are two chemicals in biosolids that can pollute groundwater and surface water, according to the author?

Sam Hadeed, "Addressing Concerns with Land Application of Biosolids," National Biosolids Partnership, August 2, 2002. www.biosolids.org. Reproduced by permission.

3. According to an investigation by independent scientists, as cited by Hadeed, how severe is the risk from organic compounds when biosolids are applied to land in typical amounts?

The most common concerns related to land application of biosolids are water quality, synthetic organic compounds, pathogens, and trace metals. The examples given below provide perspectives that can help to address concerns expressed by communities. . . .

Proper Management Protects Water Quality

Water quality protection in a land application program addresses the most likely potential source of water pollution from applying biosolids as a nutrient source: the nitrogen contained in the biosolids. Nitrate, the mobile form of nitrogen, is released slowly from organic material such as biosolids and therefore moves with soil water more slowly than does soluble nitrate from chemical fertilizers. For agricultural use, application rates are based on the calculated amount of nitrogen the crop will use, coupled with estimates of the decomposition rate of organic nitrogen contained in biosolids to forms available to crops (mineralization). Nitrogen taken up by crops is not available for leaching to groundwater or surface runoff and the basic agronomic principle of matching as closely as possible biosolids applications to the nitrogen need of the crop assures protection of groundwater.

Biosolids also are managed to prevent undesirable levels of phosphorus buildup in soils. Unlike nitrogen, such levels are not identical with the amounts taken up by crops, since phosphorus becomes less mobile over time through soil reactions. Unlike nitrogen, the soil levels at which phosphorus becomes mobile are generally much higher than the amounts needed to feed crops and those environmentally significant levels represent the appropriate limits. In addition, the phosphorus contained in biosolids is generally less soluble (bioavailable) than that contained in animal manures and chemical fertilizers, and thus less likely to cause surface water enrichment.

With respect to most trace metals potentially present in biosolids, water quality is protected by the low mobility of these pollutants in soils. The concern in this case is based on possible transfer to plants, animals and humans, since these components of biosolids remain in the upper layer (plow layer) of the soil. Applying biosolids can actually help improve water quality. The natural organic matter in biosolids helps bind soil particles to improve soil structure, making it more resistant to erosion. Adding biosolids to soil also increases the soil's ability to retain water. The same characteristics that can improve soil structure also prevent the movement of biosolids applied to the surface (for example, on pastures where biosolids are not incorporated). Analysis of runoff water from pastures has shown that surface-applied biosolids are less likely to pollute than are animal manure or chemical fertilizers.

Negligible Risk from Organic Compounds

The possible presence of organic compounds is sometimes a public concern when biosolids are applied to land. Based on nationwide U.S. EPA [Environmental Protection Agency] surveys of wastewater treatment plants, the maximum concentrations of organic pollutants potentially of concern in biosolids are below amounts found to produce a toxic effect when applied to the land. By way of comparison, when contained in biosolids applied at agronomic rates, these compounds will typically be applied in amounts 10 to 100 times lower than the amounts applied at recommended rates for agricultural pesticides. They also would have to be directly ingested to have any potential effect on consumers, since plants do not absorb them to any significant degree.

While not each and every individual organic compound potentially present in biosolids has been extensively tested on agricultural land, information on the behavior of various classes of compounds (i. e., those that are chemically similar and behave in similar ways in the environment) can be used to assess their impacts. This information can be coupled with data on the concentrations of the compounds that are of potential concern through various environmental exposure routes to establish risk. For the compounds potentially of concern in biosolids, independent scientists conducted detailed risk assessment as part of the development of the Part 503 Rule [the federal regulation governing biosolids]. When evaluated in light of actual concentrations found in biosolids, a negligible level of risk was found.

Minimizing Risk of Disease

Risk of disease (like the concern for ingesting organic priority pollutants) is negligible when treated biosolids are applied to land. Processes to reduce pathogens (typically digestion or lime stabilization) are required before any biosolids can be applied to land. These disinfecting processes do not sterilize biosolids. However, disease-causing organisms are reduced to such low levels that they do not present risk of infection or disease, especially when combined with the required crop management practices for land-applied biosolids.

If some pathogens remain in biosolids following the required stabilization processes, they die off in the soil environment, and their movement through soils is very limited. Good management practices in applying biosolids (e.g., setback distances) also serve to prevent runoff of surface water that might contain pathogens. Not only has there never been a demonstrated instance of human or animal disease resulting from biosolids that are recycled in accordance with regulations, but epidemiological evidence with real people living where biosolids are applied verifies that this practice does not represent a health risk.

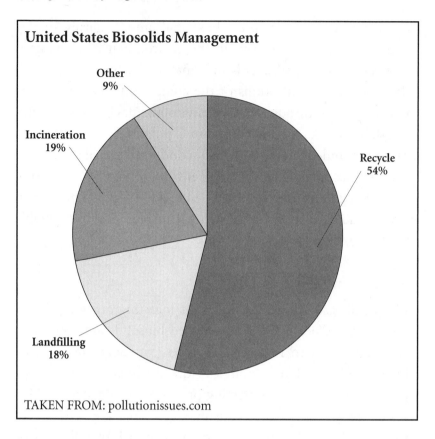

United States Biosolids Management

Other
9%

Incineration
19%

Recycle
54%

Landfilling
18%

TAKEN FROM: pollutionissues.com

Avoiding Heavy-Metal Pollution

Perhaps the best-known consideration in applying biosolids to land is the management of trace metals contained in biosolids. As mentioned in the discussion on water quality above, these metals are regulated based on well-known behaviors in soils, and to prevent uptake by crops to protect the food chain. Regulatory standards establish the amounts of trace metals that may be applied to soil to allow unrestricted future use of each site. In addition, the amounts of metals in biosolids-treated soils have been established by regulation to protect individuals that might actually consume biosolids from land application sites (by soil adherence to crops or directly).

For the Part 503 Rule U.S. EPA established management practices and numerical limits (standards) to safeguard public health and the environment by examining the probability of individuals' exposure to pollutants from biosolids. These standards were based on data provided by researchers throughout the United States. The standards also include special populations at greater risks (e.g., small children playing in gardens where biosolids have been applied). . . .

Reaching Out to Communities

As biosolids programs are becoming more publicly visible, interest in such programs is intensifying. Public meetings, field days, and media coverage all can help to provide the informational outreach needed to operate a program. These efforts are even more important given the greater likelihood that inaccurate, negative information will reach more people than in the past.

Outreach materials and ongoing efforts to achieve public acceptance should include such messages as environmental benefit, agricultural benefit and the protection of the environment, health and safety embodied in existing practices and regulation. Achieving public acceptance includes outreach efforts to the general public, and specifically developing local community support from the outset. In many cases, such support will only be forthcoming if biosolids programs are managed to go beyond the minimum (i.e., regulatory) requirements. One example of such an approach is to include voluntary water well monitoring as part of a land application program. Not only does this effort serve to reassure farmers and neighbors, the information acquired can be very useful in addressing any future questions and concerns about water quality. . . .

Addressing Public Concerns

To ensure ongoing and up-to-date scientific information is available for outreach efforts, methods for compiling data and exchanging information on beneficial use issues are critical. Research in related areas should be evaluated for its relevance to the issues surrounding biosolids management. Efforts to gain new insights and data should be accompanied by a commitment to provide this information to the community and to interested members of the general public.

Some specific areas to address in an outreach program are local concerns: aesthetics—including odors, traffic, storage, and quality assurance.

It is critical that everyone involved in biosolids projects be responsive to the underlying concerns many people have, regardless of their scientific validity. They must be particularly sensitive to the impacts of local nuisance factors that may be associated with biosolids (e.g., truck traffic and odors). Even if people's fears and concerns are not based on scientific evidence, they are nonetheless real and must be respected to get at the root of issues that can impede success and address them effectively.

Periodical Bibliography

The following articles have been selected to supplement the diverse views presented in this chapter.

Andy Coghlan — "Faeces on Food Crops—Safer than You'd Think," *New Scientist*, September 16, 2006.

Economist — "How Green Is Your Apple?" August 26, 2006.

Ecos — "Radioactive Waste Technology Is Locking Up Interest," October/November 2005.

Glen Gibbons — "Hazmat Keeps on Truckin'," *GPS World*, October 2004.

Christine Graef — "Native Dumping Ground," *E: The Environmental Magazine*, September/October 2005.

Michael J. Meyer, Waleed Abu El Ella, and Ronald M. Young — "Disposal of Old Computer Equipment," *CPA Journal*, July 2004.

Michael D. Moore et al. — "Converting Biosolids to a Renewable Fuel," *BioCycle*, October 2006.

Susan Moran — "Panning E-waste for Gold," *New York Times*, May 17, 2006.

Robin M. Nazzaro — "Low-Level Radioactive Waste: Future Waste Volumes and Disposal Options Are Uncertain." U.S. Government Accounting Office, 2004.

New Scientist — "Dangerous Dumps," July 2005.

Abby Rockefeller — "Sewers, Sewage Treatment, Sludge: Damage Without End," *New Solutions*, 2002.

Paul Vitello — "Clearing a Path from Desktop to the Recycler," *New York Times*, November 11, 2006.

Jerry Wilson — "How Boyd County Got Saved," *Nebraska Life*, January/February 2006.

OPPOSING
VIEWPOINTS®
SERIES

CHAPTER 4

Can New Technologies Solve Waste Problems?

Chapter Preface

The treatment of America's garbage has undergone rapid changes in the past two decades. The "garbage crisis" of the late 1980s prompted politicians to pass new regulations concerning the treatment of America's solid waste. In response, businesses developed new technologies and processes to meet regulatory requirements. While industry groups and supporters of private-enterprise solutions to waste disposal have hailed such developments, environmentalists are wary of the "corporatizing of trash."

American business has responded to new regulations governing garbage disposal by innovating methods of treating waste and by increasing capacity for burying it. One development has been the so-called mega-landfill. Small and municipal landfills often could not afford to install the new technology required to meet the demands of new landfill regulations passed in the early 1990s. As a result, cities began to hire large corporations, such as Waste Management International, to deal with their trash. These corporations buried the waste in extremely large landfills, often located in rural areas far removed from the cities in which the trash originated. Professor William Rathje, an expert in waste management at the University of Arizona, points out that "once it is produced, trash has to go someplace; and, in this day of heightened environmental protection, economies of scale select strongly for those mega-landfills." Contrary to what might be expected, rural communities often welcome the building of these large landfills; they provide much-needed jobs in disadvantaged areas.

In addition to the mega-landfills, corporations also developed waste-processing technologies. High-temperature incineration has been touted as a way to generate energy from rubbish, while the "bioreactor," a new type of landfill that uses chemicals and bacteria to break down buried solid wastes rapidly, is being introduced in communities around the country.

Not everyone is convinced that these new developments are the best solution to today's waste problems. Some environmentalists mistrust corporations, noting that businesses that make money with every ton of garbage they bury are unlikely to make serious attempts to reduce the waste stream through recycling. This type of privatization "envisages handing over all collection and disposal to the waste disposal industry. Here governments will not be encouraging resource recovery but waste disposal," observes the environmental advocacy group Zero Waste. Rural activists also oppose mega-landfills, believing that accepting other communities' trash reduces the chances for more meaningful economic development. "The problem with importing trash is that it is a short-term revenue fix for these rural communities. It does not get to the heart of economic woes in these counties," writes a member of the North Carolina Conservation Network.

No doubt these battles over the best way to treat our waste will continue as new technologies and policies come into being. The following viewpoints offer a variety of perspectives on the capability of waste-processing technologies to solve our garbage problem.

> "Corporations that handle much of the country's garbage today make their money in direct proportion to the amount that gets thrown away: the more trash, the more cash."

Modern Landfills Encourage Waste and Endanger the Environment

Heather Rogers

Environmental writer and filmmaker Heather Rogers believes that the consolidation of the garbage-processing industry in the United States has derailed efforts to reduce waste in our society. In the following viewpoint from the Nation *magazine, she argues that the development of mega-landfills has resulted in rural areas being burdened with waste generated by urban population centers. The large companies that control these landfills engage in "greenwashing,"—claims that they are protecting the environment; however, landfill technologies like "the bioreactor," designed to speed decomposition thereby increasing landfill capacity, harm the planet by producing increased levels of greenhouse gases.*

Heather Rogers, "Titans of Trash," *The Nation*, vol. 281, no. 21, December 19, 2005, pp. 21–23. Copyright © 2005 by *The Nation Magazine*/The Nation Company, Inc. Reproduced by permission.

As you read, consider the following questions:

1. What, according to the author, is the connection between mega-landfills and the recent spate of severe hurricanes?

2. How did increased regulation help the larger garbage-handling companies and hurt the smaller ones, according to Rogers?

3. According to the author, why don't the large waste-disposal firms invest in recycling and waste reduction?

Katrina, Rita, Stan, Wilma: The hurricanes of 2005 follow [the previous] season's onslaught of vicious storms that slammed Florida and the Caribbean. But [in 2005], the toll [was] much higher. Katrina's raging winds and water killed hundreds, displaced more than a half-million, caused tens of billions of dollars in damage and pulverized the social order of one of the largest cities in the United States. A few weeks later, Rita incited the biggest exodus ever seen in American history, as more than 2 million people fled their homes. And outside this country, Hurricane Stan triggered horrific mudslides in Guatemala that buried more than 1,000 people alive, turning their villages into mass grave sites. Today it's increasingly accepted that the ferocity of these storms is stoked by global warming.

Landfills Produce Greenhouse Gases

Think greenhouse gases, and what most readily comes to mind are industrial smokestacks and bulky, gas-gluttonous SUVs. But there is a stealthier and increasingly threatening culprit driving climate change, and it's equally close to home: the massive landfills where we stash our garbage.

In contrast to the Gulf Coast, things appear relatively calm in the Virginia countryside. Amid the encroaching suburban sprawl that rolls south from Washington [D.C.], half-vacant strip malls along Highway 3 slowly give way to verdant corn and soy fields. About twenty miles east of Fredericksburg, a neatly tiered hill, about 170 feet high, asserts itself in the hazy blue sky, well above the billowing oak trees. This is the King George County Landfill, operated by the world's chief rubbish handling corporation, the multibillion-dollar Waste Management Inc. (WMI).

King George looks like any typical modern-day garbage landfill, except it's not. Atop this grass-cloaked trash butte is a ten-acre parcel dedicated to research and development on the next generation of sanitary landfills, known in the industry as the "bioreactor." This facility pumps enormous volumes of toxic liquids into the guts of the landfill to speed the decomposition of organic materials, which will hasten the dump's settling and make more room for discards. The bioreactor's intensified decaying also forces a sharp spike in methane gas, a natural by-product of biodegradation. This is a problem because methane is a serious global warming threat: According to the Environmental Protection Agency, it's twenty-one times more heat-trapping than carbon dioxide. Peter Anderson, executive director of the Center for a Competitive Waste Industry, explains, "Bioreactors are giant greenhouse gas machines."

Regardless, firms like WMI—which, according to a senior company official, has spent more than $22 million on R&D [research and development] at the King George facility and nine other demonstration bioreactors in North America—are working hard to make this new technology an industry standard. Michael Thomas, the engineer at King George, says that if these tests go well, "you're going to see landfills all over the country converting to bioreactors."

Despite a growing chorus of environmentalists, climate experts and political leaders calling urgently for remedies to global warming, projects like the bioreactor are proceeding apace, largely outside the realm of public debate. But the bioreactor would drastically reshape the way rubbish is disposed of in the trash-rich United States, generator of more than 30 percent of municipal wastes created by Organization for Economic Co-operation and Development–member countries.

Burying Waste Means Corporate Profits

So what are the forces driving the development of this new technology? The bioreactor, with all its hazardous potential, is the latest product of an industry that's undergone intense corporatization over the past several decades, fostering a system reliant on ever greater levels of wasting, no matter the environmental toll. The corporations that handle much of the country's garbage today make their money in direct proportion to the amount that gets thrown away: the more trash, the more cash. In fact, these companies earn the highest profits from castoffs that get landfilled; burying rubbish generates more before-tax income than all other waste company operations combined. And since organic items make up almost two-thirds of all landfilled waste, these firms would stand to lose vast profits if those discards were diverted to, say, a composting program. Bioreactor technology, by contrast, is designed to maintain maximum flows of discards into the ground.

According to David Kirkpatrick, managing director of a Durham, North Carolina, firm that invests in clean technologies: "Clearly, [for-profit landfill operators] will make more money the more tons that come in. Any front-end separation for composting reduces the volumes going into the landfill, and that reduces revenues."

In an Orwellian flipping of the script, WMI markets the bioreactor as a means of "enhancing environmental protection." The corporation claims it can readily capture the dramatically increased methane output through wells dug into the bioreactor. This gas would then be channeled to power plants for conversion into electricity. In full green-washing mode, Gary Hater, the firm's senior director of bioreactor technology, lays it out: "By going from methane to energy, you're decreasing reliance on imported fuels and domestic fuels; you're taking a waste material and generating green energy in a renewable way—you're creating green energy!"

Never mind all the energy that's wasted when discards are buried in landfills instead of getting reused or recycled. According to Neil Seldman of the Institute for Local Self-Reliance, 80 percent of U.S. products are used once, then thrown away. If those items were built to last longer, if they were easier to repair and reuse, a lot less energy would be consumed in the first place.

Consolidation of the Waste-Processing Industry

How is it that waste treatment—such an integral part of daily life and environmental health—is overseen by a handful of corporations that can ignore these realities? As of 2003, the country's top three garbage firms controlled more than 40 percent of the almost $45 billion market. And beneath their greenwashed veneers, these firms are mostly concerned with revenues. Their dominance has come about thanks to a dramatic industry restructuring in the past thirty years in which garbage conglomerates took over the waste business, rolling up mom-and-pop outfits and shoving publicly owned operations to the sidelines.

Two early players in the corporatization of garbage were WMI and Browning-Ferns Industries [BFI]. These giants spearheaded the sector's consolidation, buying out smaller firms in towns across the Sunbelt, then spreading north and into international markets. In building their empires, WMI and BFI also took their trash public, offering stocks to raise revenues, diverging sharply from previous industry norms. And in the 1990s a handful of newly powerful corporations like Allied Waste Industries and USA Waste Services followed WMI and BFI into the staggeringly profitable world of high-volume trash.

© CartoonStock.com

Regulation Favors Large Companies

Viewed from the outside, this momentum appeared to hit a snag when, in 1991, the EPA [Environmental Protection Agency] instituted tighter controls for landfills. While further regulation may have seemed like a good move for the environment and an obstacle for the garbage corporations, in reality the big players benefited. The measure, known as Subtitle D, required all landfills to protect against groundwater and air pollution by converting to "dry tomb" technology; this entailed the costly installation of liners, along with gas and liquid leachate collection and monitoring systems. The big trash firms liked Subtitle D because it created barriers to entry for their less capitalized, smaller rivals and would price out many municipalities, facilitating yet further consolidation of the waste industry.

With their competitors on the ropes, the conglomerates went on a buying spree, plucking up defunct dumps to upgrade and opening new disposal sites in regional hubs like Pennsylvania, Virginia, Ohio and Michigan. Here the trash corporations built mammoth new "mega-fills," giant disposal sites fully outfitted with all the required monitoring equipment.

Waste Exported to Rural Areas

During this time, the overall number of landfills nationwide declined as substandard dumps were shuttered. But meanwhile, under the conglomerates, actual burial capacity soared. Whereas yesterday's landfills could take in anywhere from dozens to several hundred tons of refuse daily, new mega-fills could handle thousands of tons per day. This facilitated another shift in the industry, toward the inequitable exporting of wastes across state lines, mostly from urban centers like New York City to economically hollowed-out rural regions nearby.

Under the current regime, corporations are in charge of treating a huge portion of US household discards. With their colossal budgets and the political power they therefore wield, the trash giants exercise considerable influence over the way garbage will be treated in the coming decades. And they aren't investing in waste-reduction, recycling or composting technologies in any significant way, although these are proven, ecologically sound methods. The bioreactor, with its dangerous greenhouse gas emissions, represents the latest effort by these firms to keep the spent, used and broken riches of our society pouring into the landfill.

Eliminating obstacles to accumulation has long been the goal of US business—whether it's manufacturing or waste disposal—and has led to a profoundly antienvironmental, undemocratic system. With increasing corporate ownership of garbage treatment and disposal, the public has been denied a meaningful role in devising better ways to reduce and recycle our wastes. Concern about profits and expanding markets, not human and environmental health, is what drives how we handle our garbage today.

| "Modern landfills are specifically designed to protect human health and the environment by controlling water and air emissions."

Modern Landfills Are Safe for the Environment

National Solid Wastes Management Association

In the following viewpoint, excerpted from a report published by the waste industry group the National Solid Wastes Management Association, data about improvements in landfill technology over the last thirty years is presented. Landfills once were truly serious dangers to the environment; however, by installing leachate containment systems to trap liquids percolating from landfills and by developing methods to control methane and other gas emissions, environmental engineers have minimized the impact of landfills on the environment, the author argues. In doing so they have minimized the risks to human health, the association maintains.

As you read, consider the following questions:

1. What were two environmentally hazardous practices in older landfills, according to the author?

National Solid Wastes Management Association, "Modern Landfills: A Far Cry from the Past," March 2006. www.nswma.org. Reproduced by permission.

2. As explained by the association, how do modern land-fills contain and treat liquids percolating through the refuse?

3. How does a "bioreactor" work, as described by the author?

In less than 30 years, landfills changed from little more then holes in the ground to highly engineered, state-of-the-art containment systems requiring large capital expenditures. Typically, older landfills were designed by excavating a hole or trench, filling the excavation with trash, and covering the trash with soil. In most instances, the waste was placed directly on the underlying soils without a barrier or containment layer (liner) that prevented leachate (water percolating through the waste and picking up contaminants) from moving out of the landfill and contaminating groundwater.

Dumped garbage was openly burned to save space for future waste disposal, creating air pollution and health hazards. When the waste reached a predetermined height, a final cover of soil was placed on top and sometimes vegetation was planted. In many cases, the vegetation failed to grow or died because of methane gas (a natural byproduct of waste degradation) escaping through the final cover. Also, the landfill gas could move offsite into buildings and homes potentially creating explosion risks.

In contrast, modern landfills are specifically designed to protect human health and the environment by controlling water and air emissions. . . .

Building a Modern Landfill

Liquid containment within a modern landfill results from a combination of the liner and the leachate collection system performing complementary functions to prevent groundwater contamination. Liners prevent leachate and gas migration out of the landfill while directing liquids to the leachate collection system.

Liner systems are typically constructed with layers of low-permeability, natural materials (compacted clay) and/or synthetic materials (high-density polyethylene). The leachate collection system removes the liquid contained by the liner. A typical leachate collection system may consist of (from bottom to top) a perforated leachate collection pipe placed in a drainage layer (gravel), a filter blanket, and a leachate collection layer.

Waste is placed directly above the leachate collection system in layers. Delivered waste is placed on the working face that is maintained as small as possible to control odors and vectors [natural factors that might spread refuse-generated toxins]. Heavy steel-wheeled compactors move the waste into the working face to reduce the waste's volume. At the end of each day, the waste is covered with six inches of soil or an alternative daily cover (foam, tarps, incinerator ash, compost) to control vectors, odors, fires, and blowing litter.

Once the landfill has reached its permitted height, the landfill is closed and engineered to prevent water infiltration by installing a low permeability cap similar to the liner system. The final cap can be comprised of a compacted clay and/or a synthetic material. A granular drainage layer is placed on top of the low-permeability barrier layer to divert water off the top of the landfill. A protective soil cover is placed on top of the filter blanket and topsoil is placed as the final layer to support vegetation.

In short, the sophisticated engineered systems in a modern landfill ensure protection of human health and the environment by containing leachate that can contaminate groundwater, preventing the infiltration of precipitation that generates leachate after closure of the landfill, and collecting landfill gas which can be used as an energy source or destroyed.

Controlling Greenhouse Gases

Historically, open dumps were associated with a number of environmental problems; however, the goal of today's modern landfill design and operation is to control and eliminate any environmental impacts.

When MSW [municipal solid waste] is disposed of in a landfill, naturally occurring microorganisms (bacteria) degrade the waste. The amount of water in and the temperature of the MSW control the rate of degradation. This process turns the organic portion of the waste into methane (a primary constituent of natural gas) and carbon dioxide in about equal proportions. The degradation process also generates very small quantities of organic compounds.

Additionally, some organic compounds may be released directly into the gas from products contained in the waste, such as cleaning products. The non-methane organic compounds in the gas amount to less than one percent of the total gas created by waste degradation. Gas generated can threaten human health and the environment if it migrates off site or is not collected and destroyed.

Under federal Clean Air Act standards, larger modern landfills with estimated uncontrolled emissions of 55 tons per year of NMOC [non-methane organic compounds] or more are required to install a gas collection and destruction system. Large landfills are defined as having a design capacity of equal to or greater than 2.76 million tons and 3.27 million cubic yards. Many smaller modern landfills voluntarily install gas collection and destruction systems for various reasons, including earning emission reduction credits by reducing their greenhouse gas (GHG) emissions or being a "good neighbor." The gas collection system directs gas to a central location where it can be processed and treated depending on its ultimate use. From this point, gas can be destroyed in a flare (similar to a gas stove) or used as an energy source to produce electricity, replace natural gas, or as a fuel to power vehicles.

Fewer Emissions, More Green Energy

According to EPA [Environmental Protection Agency] studies, modern landfills generate significantly lower concentrations of non-methane organic compounds then older sites. Of 42 non-methane organic compounds regulated by federal clean air rules, 59 percent of them are one to three orders of magnitude lower (an order of magnitude is a ten fold decrease) in concentration in modern landfills than in older landfills. In fact, 3 of 24 non-methane organic compounds were not detected at modern landfills, but were present at high concentrations in older landfills. Since comparable data are from test programs conducted in the late 1980s, older landfills are likely to have higher concentrations of non-methane organic compounds than modern landfills.

In addition to the environmental benefits of lower concentrations of non-methane organic compounds in gas from modern landfills compared to older ones, the potential risks to human health and the environment of gases from modern landfills is significantly less than older landfills because the devices used to combust the gas have destruction efficiencies of more than 99 percent for methane and greater than 98 percent for all other non-methane organic compounds.

As noted earlier, collected landfill gas can be used to generate electricity or heat for powering industrial facilities, providing lighting and temperature control to homes and businesses, or as a fuel for use in vehicles. EPA's latest data show that there are more than 375 operational gas-to-energy projects in 38 states. These projects collect some 74 billion cubic feet of landfill gas and generate 9 billion kilowatt hours of electricity per year.

Because methane is a potent GHG (approximately 23 times more global warming potential than carbon dioxide), an additional benefit of modern landfill gas collection and destruction equipment is the reduction of methane released to the atmosphere where it contributes to global warming. Recent

The Three Different Types of Bioreactor Landfill Configuration

- **Aerobic**—In an aerobic bioreactor landfill, leachate is removed from the bottom layer, piped to liquids storage tanks, and re-circulated into the landfill in a controlled manner. Air is injected into the waste mass, using vertical or horizontal wells, to promote aerobic activity and accelerate waste stabilization.

- **Anaerobic**—In an anaerobic bioreactor landfill, moisture is added to the waste mass in the form of re-circulated leachate and other sources to obtain optimal moisture levels. Biodegradation occurs in the absence of oxygen (anaerobically) and produces landfill gas. Landfill gas, primarily methane, can be captured to minimize greenhouse gas emissions and for energy projects.

- **Hybrid** (Aerobic-Anaerobic)—The hybrid bioreactor landfill accelerates waste degradation by employing a sequential aerobic-anaerobic treatment to rapidly degrade organics in the upper sections of the landfill and collect gas from lower sections. Operation as a hybrid results in the earlier onset of methanogenesis compared to aerobic landfills.

Environmental Protection Agency.

studies using EPA's lifecycle model have shown that methane emissions from modern landfills declined 54 percent from 1970 to 2003 because of increased use of landfill gas collection and control systems. . . .

Preventing Groundwater Contamination

As referenced earlier, few older landfills had liners and leachate collection systems to prevent leachate movement out of the landfill. Modern landfills are equipped with liners and leachate collection systems that prevent the leachate from leaving the facility and contaminating groundwater. Based on recent EPA studies, a liner and leachate collection system constructed to current standards typically has a liquid removal efficiency of 99 to 100 percent and frequently exceeds 99.99 percent.

EPA research shows that most trace chemicals are detected at lower concentrations in leachate from modern landfills than from older ones. In most cases, contaminant concentrations in leachate from modern landfills are one to two orders of magnitude less compared to older landfills. Moreover, EPA research anticipates that the quality of leachate will continue to show improvement over time as the existing public database for modern landfills increases.

Releases of trace constituents contained in the leachate from modern landfills are practically eliminated because leachate is collected, removed, and treated. Leachate collected at landfills is either treated on-site or transported off-site for treatment. Federal requirements mandate that treatment must meet drinking water quality standards, which are set to prevent harm to public health, or more stringent state standards to protect sensitive environments (high quality streams, trout streams).

Research has shown that leachate treatment facilities at modern landfills are capable of removing 100 percent of the trace organics and over 85 percent of the heavy metals.

To ensure the liner and leachate collection system are operating properly, groundwater-monitoring wells are installed around the landfill and tested regularly for indications of releases from the landfill. Groundwater quality reports are supplied to the appropriate state regulatory authority on a routine basis. If contaminants indicative of a release from the landfill are found in monitoring wells at levels above health-based standards, the landfill must correct any problems that resulted from the release and restore the groundwater to its original quality. . . .

Bioreactors and Biocovers

The waste industry continues to investigate innovative operations and designs that further protect human health and the environment. One promising innovation is the bioreactor landfill. A bioreactor landfill operation and design adds liquids and/or air to the waste, which accelerates the waste biodegradation process and waste stabilization.

Based on research, the environmental benefits of a bioreactor landfill include:

- Shorter time periods (7–10 years) over which air and water emissions are generated compared to 30 or more years in a conventional landfill;

- Shorter post-closure care periods (10–15 years) compared to 30 or more years for a conventional landfill;

- Increased efficiency of the gas collection system; and

- Quicker return of the property to a productive end-use.

Another promising innovation is the use of biocovers (composted yard waste used as final cover) to further reduce air emissions at landfills. The benefits of biocovers are that air emissions of methane and other organic compounds are oxidized and destroyed in the biologically active compost. Research has shown that biocovers are effective for controlling air emissions when used:

- On areas where more waste will be added at a later date and a landfill gas system is not fully operational; or

- To control air emissions when the gas system is shutdown for maintenance and repair.

Modern Landfills Minimize Environmental Harm

As in the past, landfills will continue to play an important role in our nation's MSW management system. However, gone are the past problems associated with older landfills such as groundwater and air contamination, acceptance of hazardous waste, and inappropriate locations in sensitive areas. Modern landfills, in contrast, are highly engineered containment systems that are designed and operated to minimize the impacts of municipal solid waste disposal on human health and the environment.

"High temperature incineration will continue to play an important role in the future for the safe and effective treatment of much of the organic hazardous wastes that will continue to be generated by United States industry."

Modern Incinerators Can Safely Dispose of Waste

Environmental Technology Council

The following viewpoint by the Environment of Technology Council, the trade association of the waste disposal industry, makes the case that incineration is the best technology available to deal with hazardous organic waste. By explaining how incinerators operate, how their processes are monitored, and how the residues from incineration are dealt with, the author makes the case that these facilities for high-tempature burning of waste are the only feasible alternative for destroying harmful chemicals generated by U.S. industry.

As you read, consider the following questions:

1. How many tons of hazardous waste are generated annually, according to the council?

Environmental Technology Council, "Hazardous Waste Incineration: Advanced Technology to Protect the Environment," February 2007. www.etc.org. Reproduced by permission.

2. As described by the author, how does an incinerator destroy organic waste?

3. What are the three steps involved in processing ash and residue left over from incineration, as explained by the council?

In our society, most industrial processes produce waste which the U.S. Environmental Protection Agency (EPA), acting under the Resource Conservation and Recovery Act (RCRA), has determined can be detrimental to public health or the environment if not properly managed.

Hazardous wastes are generated, for example, in the production of cosmetics, pharmaceuticals, detergents, household paint and cleaning products, light bulbs, telephones, televisions, newspapers, garden pesticides, computers, chemicals, gasoline, and even automotive safety devices such as air bags.

Hazardous Wastes Continue to Be Generated

Companies have made considerable strides in recent years to reduce or recycle hazardous wastes from their production processes. They have been encouraged to do so by public policies and by their own economic interests in reducing their cost of waste disposal. The high cost of incineration, relative to other forms of treatment or disposal, has been an incentive to reduce the generation of incinerable wastes.

Nevertheless, more than 38 million tons of hazardous wastes are now generated annually, and large amounts of hazardous waste will continue to be generated in the future even under optimistic but realistic expectations.

Moreover, many industrial wastes which are not yet defined as hazardous should be brought under hazardous waste regulation in the future. For example, there are over 30,000 government-registered pesticides that contain toxic and hazardous constituents, of which fewer than 20 are regulated now as hazardous waste when discarded. Future federal regulations should define more of these and other wastes as hazardous, as many states presently do.

In light of the risk these unregulated wastes pose, many responsible companies presently send their non-regulated industrial wastes to commercial incineration facilities for safe and proper treatment.

Another major area of need for waste treatment and disposal is the cleanup of contaminated sites. The country's worst environmental sites are listed in the National Priority List under the Superfund program. The toxic materials from these sites should be properly treated and disposed. . . . As many as 3,000 other contaminated sites will need to be addressed under the RCRA, through the corrective action program. A top priority in this country's environmental program must be the cleanup of problem sites that pose a risk to human health and the environment.

Best Technology Available

In order to treat these wastes, destruction in high temperature incinerators has been determined by EPA, after extensive expert and public review, to be the Best Demonstrated Available Technology (BDAT) for most organic hazardous wastes. This is because incineration safely and effectively destroys the hazardous constituents in the waste, as discussed in the subsequent sections of this report.

In the U.S., modern hazardous waste incineration is a widespread technology. Most hazardous waste incinerators . . . are owned and operated by the factory or other facility that generates the waste, and are located on the generating site. Fewer than 25 incinerators that accept off-site generated wastes (i.e., "commercial" incinerators) serve small businesses and other generators who cannot effectively or economically incinerate their hazardous wastes on-site. Today, 95% of hazardous waste generators, including many small businesses, depend entirely on off-site facilities for management of *all* of their hazardous wastes.

In short, high temperature incineration will continue to play an important role in the future for the safe and effective treatment of much of the organic hazardous wastes that will continue to be generated by U.S. industry. It is also a necessary component, as noted above, of the cleanup of organic wastes at thousands of existing Superfund and other remedial sites.

Incinerator Operation

A typical hazardous waste incinerator consists of a rotary kiln (primary combustion chamber), an afterburner (secondary combustion chamber), connected to an air pollution control system, all of which are controlled and monitored.

Rotary kilns Both solid and liquid wastes are introduced into the rotary kiln, in which the temperature is typically above 1800°F. Temperature is maintained at this level by using the heat content of the liquid wastes or by introducing supplemental fuels into the chamber, such as natural gas.

Liquid wastes generally are pumped into the kiln through nozzles, which atomize the liquids into fine droplets—as small as one microgram (one millionth of a gram)—for optimal combustion. Solid wastes may be fed into the kiln in bulk or in containers, using either a conveyer or a gravity feed system.

The kiln slowly rotates so that the solid wastes are tumbled, to assure that they are exposed on all sides to the high temperature in the kiln, much as the rotation of a clothes dryer maximizes the exposure of the clothes to the hot air in the dryer. A large fan draws excess air (containing oxygen) into the system to increase combustion efficiency.

The flame and high temperature in the kiln cause the organic and some of the metal wastes to be converted from solids or liquids into hot gases. These hot gases pass into the afterburner. Any inorganic materials (metals, such as zinc or lead) that have not been converted into gases drop out as ash at the end of the kiln, into a container, for further management.

Afterburner Atomized liquid wastes and/or supplemental fuel are injected into the afterburner, where temperatures are typically maintained at 2200°F. or higher. These atomized liquids and the hot gases entering the afterburner from the kiln are mixed with air and passed through the hot flame in the afterburner. The heat and flame break down the chemical bonds of the gaseous and atomized organic compounds into atoms. These atoms recombine with oxygen from the air in the chamber to form stable compounds primarily composed of nonhazardous chemicals such as carbon dioxide and water (*i.e., steam*).

Air pollution control system (APCS) The gases exiting the secondary chamber are cooled and cleaned in the APCS. The APCS removes particulates (small solid matter) and the remaining hazardous constituents—such as metals which were not destroyed by the incineration process—down to levels established as safe by the regulations and the facility's permits. . . .

Controls and monitoring Operation within the key parameters of the combustion process are assured by systems of monitors and computer controls. These systems make automatic adjustments to key functions as necessary. For example, if temperatures begin to drop below desired levels, supplemental waste fuels are automatically injected. Conversely, if temperatures rise above the desired range, waste feeds are reduced.

All regulated incinerators have waste feed cut-offs (WFCOs) to assure protective operations. WFCOs automatically stop the feeding of waste into the incinerator if any of the key parameters even momentarily falls outside the narrow range of operating requirements.

There is also continuous monitoring and recording of key indicators, so that a permanent record is maintained, verifying operation of the incinerator within these parameters. Frequently, as many as twenty separate parameters are monitored and recorded.

Dealing with Ash and Residue

Residue management The rotary kiln discharges an inorganic ash into a large container. The ash, and any residue from the APCS, is analyzed to assure that it does not contain any hazardous organic constituents above concentration levels specified in EPA's regulations as safe for land disposal. These concentration levels are almost always less than one part per million for any organic hazardous constituent.

This inorganic residue is further treated by mixing it with chemical stabilizers to chemically bind the constituents. The chemically stabilized inorganic residue is analyzed to assure that the metals cannot leach out of the residue above the low levels specified in EPA's rules. The facility retains the results of its analyses, and must certify that the residue meets all required treatment standards.

Finally, the stabilized and certified inorganic waste residue, which now meets all required treatment standards, is placed in a hazardous waste landfill meeting EPA's Minimum Technology Requirements (MTR). MTR landfills have two liners, with a leachate collection system between the two liners. Groundwater monitoring is also provided, outside the landfill, to supplement the protection provided by the double-liner, leachate collection and leak detection systems of the MTR landfill itself.

Conclusion

Today's hazardous waste incinerators, operating under EPA and state regulations, are high-technology devices, carefully designed, controlled and maintained to assure (1) safe destruction of all hazardous organic constituents in the waste; (2) control of emissions to safe levels, generally substantially below those met by manufacturing and other industries; and (3) the proper treatment and safe disposal of any residues.

> *"[Incineration] alternatives cost a fraction of the cost of incineration, employ many more workers than incineration, and pollute far less."*

Recycling Beats Incineration in the Third World

Brenda Platt

Brenda Platt, author of several reports on waste reduction and codirector of the Institute for Local Self-Reliance, makes the case against the use of incinerators in the developing world. She argues in this viewpoint that incinerators are expensive to operate and leave residual pollution. In contrast, workers in many poorer countries already collect scrap metal, plastics, and other rubbish for reuse or recycling. With a small amount of organization and land, these workers can be made more efficient, helping developing countries to deal with their growing waste-disposal problems without the expense and residual pollution of incinerators.

As you read, consider the following questions:

1. What are some of the problems with incinerators? Do they fully eliminate the need for landfills?

Brenda Platt, "Resources Up in Flames: The Economic Pitfalls of Incineration Versus a Zero Waste Approach in the Global South," Global Alliance for Incinerator Alternatives, 2004. Reproduced by permission.

2. What is the author's estimate of the potential percentage of waste diverted by recycling and composting?

3. What are two requirements to help integrate informal recycling into a comprehensive waste management program?

The amount of unwanted discards thrown away in industrializing nations has reached crisis proportions in recent years. Rising population, rural to urban migration, increased globalization of Western consumer patterns and the proliferation of single-use disposable products and packaging are partly to blame. Landfills, typically nothing more than open dumps, are filling up and people are sprawling beyond city borders, limiting the ability to develop new landfills. In an effort to find new solutions to growing disposal headaches, many nations are shifting to the formal private sector, embracing technology—driven approaches, and turning to the old technique of waste incineration. However, incinerators—no matter where they are built—have numerous liabilities. Waste incinerators:

- generate pollution,

- harm public health,

- place huge financial burdens on host communities,

- drain local communities of financial resources,

- waste energy and materials,

- thwart local economic development,

- undermine waste prevention and rational approaches to discard management,

- have an operating experience in industrialized countries checkered with problems,

- often exceed air pollution standards,

- create toxic ash,

- can go financially bankrupt from tonnage shortfalls, and

- often leave citizens and taxpayers paying the bill.

Non-Incineration Alternatives Better for Industrializing Countries

Incineration technology, designed and tested for the discard streams and infrastructure in industrialized nations, can be expected to perform even more poorly in industrializing countries due to differences in discard stream characteristics, inadequate regulatory structures and institutional arrangements, lack of convertible currency for purchase of spare parts, lack of skilled workers, and economic systems that favor labor over capital.

Incinerator proposals—along with proposals to centralize and privatize waste management systems—are often presented as the only solution to handle growing amounts of discards. Fortunately other options exist. Indeed, non-incineration alternatives can be comprehensive, handle discarded materials from large urban areas, and be carried out in industrializing countries with minimal resources. Furthermore, alternatives cost a fraction of the cost of incineration, employ many more workers than incineration, and pollute far less. In industrializing countries, source-separation recycling and composting programs (in which recyclable and organic materials are segregated at the household level) have the potential to divert 90% of household waste from disposal, a level incineration cannot achieve.

Recycling Versus Incineration in India

Chennai (formerly Madras), India, makes a good case to illustrate the benefits of a recycling/composting approach compared to reliance on incineration. A US$41 million incinerator has been proposed for the city (population 4.3 million) that would gasify 600 tonnes per day of municipal discards. Local authorities are moving toward privatizing waste collection and, as a result, have already jeopardized community-based recycling and composting initiatives. In fact, Chennai is home to Exnora International, a nonprofit organization spearheading a decentralized recycling/composting approach that has inspired similar projects across India.

In Chennai the infrastructure exists to collect only 2,500 of the 3,500 tonnes of discards generated each day. Almost 30% is left uncollected littering streets and neighborhoods. This is typical of less-industrialized nations. Thus incinerators in Chennai, at most, could hope to receive 2,500 tonnes per day. But not all material discarded is incinerable; about 5 to 10% is considered "by-pass" materials that might, for instance, include large nonburnable items such as engine blocks, or represent waste landfilled when the incinerator is not working. In addition, on average 25% by weight of what is burned ends up as ash that still requires landfill disposal. In our Chennai example, incineration would only divert 1,750 metric tonnes a day or half of the total waste generated. In contrast, Exnora's decentralized community-based waste reduction approach involving segregated collection of recyclables and organics for composting has the potential to divert 90% of all the 3,500 tonnes generated each day. The heart of Exnora's program is teaching citizens to take responsibility for their discards and not to litter. This approach can go even further when combined with clean production policies and practices to actually design out of the system that fraction of the discard stream that cannot be safely composted, reused, or recycled. In terms of costs, the recycling/composting approach is far more cost-effective (US$4.6 million compared to US$119 million). Furthermore, the incineration system has a far more detrimental impact on the environment, local economic development, and other quality-of-life aspects such as truck traffic.

Local Knowledge Key to Recycling Success

While the figures above are theoretical, they are based on actual data of operating projects. Indeed, numerous projects around the world have demonstrated that integrated programs for waste prevention, reuse, recycling, and composting can significantly reduce disposal at a lower cost than incineration.

To be effective, discard management systems must be based on appropriate technical solutions and be designed with local conditions and needs in mind. Most industrializing countries have limited experience with operating and maintaining centralized discard handling systems. Thus, the less complicated the technology, the more successful it will be. Most industrializing countries have a significant informal sector already engaged in extensive recycling activities. A system designed in partnership with this sector and with other community efforts and micro-enterprises will also have a better chance of success. In fact, integrating the informal sector and community initiatives into citywide discard management planning is not only possible but may be the key to success. The informal sector and community programs may need only an institutional structure and land for activities such as composting to be scaleable to city levels. Indeed, community projects can become mainstream solutions. They need not be forever relegated to local small efforts.

Examples of Community-Based Waste Reduction

Some successful innovative approaches to managing discards and reducing waste in the global South include the following:

Cairo, Egypt: informal sector workers—known as *zabbaleen*—collect one-third of Cairo's household discards, about 998,400 tonnes per year. The *zabbaleen*, who live in five neighborhoods surrounding Cairo, recycle and compost 80 to 90% of what they collect. One neighborhood, Mokattam, is home to approximately 700 garbage collecting enterprises, 80 intermediary traders, and 228 small-scale recycling industries.

Mumbai, India (formerly known as Bombay): citizens have set up neighborhood associations—each known as an Advanced Locality Management (ALM)—in which members keep their environment clean and separate their discards into biodegradable and non-biodegradable types for composting and recycling. Many ALMs vermicompost (worm compost) wet organic materials and work with ragpickers to recycle other discards. About 650 ALMs exist, representing about 300,000 citizens.

Barangay Sun Valley, the Philippines: approximately 3,000 households participate in a recycling and composting program that diverts 70% of their household discards from disposal. "Biomen" collect segregated organic material (kitchen scrap and garden trimmings) for composting on a daily basis using pedicabs. The same pedicabs collect segregated recyclables from households. They deliver recyclables to the nearest "eco-shed" for further sorting and baling. Processed material is sold directly to scrap or "junk shop" dealers.

Rio de Janeiro, Brazil: in 2000, this state passed a mandatory packaging take-back law, which requires the take-back of all plastic packaging and its subsequent reuse or recycling.

Reducing Waste Is a Long-Term Project

A growing zero waste movement is gaining momentum world-wide and innovative regulatory systems requiring "extended producer responsibility" for products promise to reduce disposal even further. Local, national, regional, and international networks of concerned citizens and professionals have formed to halt proposals for new incinerators, phase out old ones, and push for alternative systems based on sustainable production and consumption patterns.

Zero waste is a worthwhile goal, but it will take some time to achieve it. Just as a journey of a thousand miles begins with a single step, so too does aiming for zero waste. The road to zero waste can begin with the simple and relatively inexpensive act of keeping organic and putrescible material out of landfills and dumps. This alone won't provide a total solution, but will go a long way toward solving problems related to dirty, leaking, and overflowing dumpsites. This is especially true in the global South where organic material makes up the largest component of the discard stream. Composting can cut the discard stream by almost half in a relatively short time period. The beauty of composting is that it can be accomplished inexpensively via low-tech means on a small-scale. More often than not, it can be done with local know-how and local resources. Keeping materials segregated is essential to success.

Periodical Bibliography

The following articles have been selected to supplement the diverse views presented in this chapter.

Christine A. Bevc et al.   "Environmental Justice and Toxic Exposure: Toward a Spatial Model of Physical Health and Psychological Well-Being," *Social Science Research*, March 2007.

BioCycle   "Saving Millions in Landfilling Costs," July 2004.

Lie-giang Chen et al.   "Effective Incineration Technology with a New-Type Rotary Waste Incinerator," *Journal of Environmental Sciences*, November 2003.

George Crawford and Richard Lewis   "Exceeding Expectations," *Civil Engineering*, January 2004.

Esther D'Amico   "Moving Through Unfriendly Territory," *Chemical Week*, January 25, 2006.

Jerome Goldstein   "Sustainable Water Supplies with Wastewater Recycling," *BioCycle*, January 2006.

Brett Hansen   "Egg-Shaped Digesters Will Be World's Largest," *Civil Engineering*, May 2006.

Jannett Highfill and Michael McAsey   "Gains and Losses from Transfers of Solid Waste," *International Advances in Economic Research*, May 2004.

Patrick di Justo   "Blue-Green Acres," *Scientific American*, September 2005.

Maria Noël Mandile   "They're Not Going to Take It Anymore," *Natural Health*, September 2002.

Waste News   "Global Alliance Blasts Burning Trash," July 21, 2003.

Monica Wilson   "Dousing the Flames," *Multinational Monitor*, January/February 2004.

For Further Discussion

Chapter 1

1. Helen Spiegelman and Bill Sheehan argue that we are producing too much waste from packaging and obsolete consumer products. Who, according to these writers, should bear responsibility for this waste? After reading the Bjørn Lomborg viewpoint, do you think he would consider waste generation in the United States a problem? Using quotations from the text, answer why or why not.

2. Heather Rogers believes that mandatory recycling can promote environmental thinking. After reading her viewpoint and the following one by Elizabeth Royte, do you think Royte would agree? According to Royte, how do the messages we receive from businesses via advertising influence the amount of waste we produce? Does Royte foresee a change? Cite evidence from the text to support your answer.

3. Eileen Gauna and Sheila Foster make the case that minority and/or poor people are more likely to live in areas with dumps, incinerators, or other environmental hazards. Does this necessarily mean that those facilities are intentionally placed in minority or poor neighborhoods? What are some alternative explanations for the presence of waste dumps in poor neighborhoods?

Chapter 2

1. According to Kivi Leroux Miller, municipalities can benefit from recycling programs; however, cities have had to adapt their programs. Name three specific adaptations made by three different cities in order to improve the efficiency of their recycling efforts.

2. According to the economic view promoted in the viewpoint by Asa Janney, does curbside recycling make sense if it must be subsidized by government? Miller cites a study that shows that recycling employs 1.1 million people nationwide. Do you think Janney believes that this is the best use of the United States' labor resources? Why or why not?

3. Asa Janney believes that sorting garbage is not a worthwhile use of his time. Can you think of any evidence from Richard C. Porter's cost-benefit analysis that a majority of Ann Arbor residents do think sorting garbage for recycling is a worthwhile use of their time? Ann Arbor is the home of the University of Michigan. Do you think that the population of Michigan at large would exhibit the same level of support for recycling as do the Ann Arborites? Why or why not?

Chapter 3

1. Morgan O'Rourke's viewpoint indicates that e-waste is a significant new environmental threat. State two facts from Dana Joel Gattuso's viewpoint that support Gattuso's belief that the dangers cited by O'Rourke are exaggerated. How might O'Rourke answer Gattuso's charge that the e-waste crisis is exaggerated?

2. Sam Hadeed claims that sewage sludge presents little or no risk to the public as a land treatment or fertilizer when properly applied. What fact from Caroline Snyder's viewpoint casts doubt on Hadeed's ability to predict that all sewage sludge is safe? Do you think that sewage sludge from different areas in the United States contains the same types and levels of pollutants? Explain your answer.

Chapter 4

1. Why does Heather Rogers believe that the "corporatization" of waste leads to more waste? Do waste disposal

companies have an incentive to reduce the waste stream? Rogers's viewpoint specifically mentions methane from landfills as a climate-harming greenhouse gas. How, according to the National Solid Wastes Management Association's viewpoint, is this problem being dealt with?

Organizations to Contact

The editors have compiled the following list of organizations concerned with the issues debated in this book. The descriptions are derived from materials provided by the organizations. All have publications or information available for interested readers. The list was compiled on the date of publication of the present volume; the information provided here may change. Be aware that many organizations take several weeks or longer to respond to inquiries, so allow as much time as possible.

Competitive Enterprise Institute
1001 Connecticut Ave. NW, Suite 1250
Washington, DC 20036
(202) 331-1010 • fax: (202) 331-0640
e-mail: info@cei.org
Web site: www.cei.org

The Competitive Enterprise Institute is a free market-oriented think tank that promotes environmental protection through the use of market mechanisms. It publishes books such as Angela Logomasini and David Riggs's *The Environmental Source* (2004) and numerous articles, including "Does Manure Make a Farm a Superfund Site?" by Steven J. Milloy.

Copenhagen Consensus Center
Copenhagen Business School, Frederiksberg DK-2000
  Denmark
+45 3815 2254
e-mail: info.ccc@cbs.dk
Web site: www.copenhagenconsensus.com

This think tank was put together by controversial scholar Bjørn Lomborg with the goal of identifying the most cost-effective ways of helping the global environment. The group's recommendations have generally been middle-of-the-road, favoring less costly actions to improve or protect mankind's natural surroundings. It publishes a series of Challenge Papers, including the "Challenge Paper on Sanitation and Water," which deals with issues of garbage and waste. Its Web site also includes a multimedia presentation, "Q&A's," about environmental issues and teaching materials on the environment.

Environmental Protection Agency, Office of Solid Waste and Emergency Response
Mailstop 5305P, 1200 Pennsylvania Ave. NW
Washington, DC 20460
(888) 372-7341
Web site: www.epa.gov/oswer

The Office of Solid Waste and Emergency Response at the Environmental Protection Agency is the federal agency responsible for regulating solid waste and wastewater in the United States. The agency publishes reports and data, such as the *Toxics Release Inventory*, concerning waste disposal.

Foundation for Research on Economics and the Environment (FREE)
662 Ferguson Rd., Bozeman, MT 59718
(406) 585-1776 • fax: (406) 585-3000
e-mail: jbaden@free-eco.org
Web site: www.free-eco.org

FREE applies economics and scientific analysis to generate and explore innovative solutions to environmental problems. It is generally oriented toward less direct government involvement in environmental issues, including the disposal of waste. Its publications include a monthly newletter (available in electronic form on the group's Web site) as well as the books *Managing the Commons* and *The Next West: Public Lands, Community, and Economy in the American West*, edited by the organization's principal scholar, John A. Baden.

Inform, Inc.
5 Hanover Square, 19th Fl., New York, NY 10004-2638
(212) 361-2400 • fax: (212) 361-2412

e-mail: calderone@informinc.org
Web site: www.informinc.org

Inform, Inc. is an independent research organization that examines the effects of business practices on the enviroment. It identifies ways to promote environmentally sound, sustainable business practices. Inform's publications include "Toxic Chemicals and Human Health" and "Promoting Waste Prevention and Design of Less Wasteful Products," both available on the organization's Web site.

National Safety Council
1121 Spring Lake Dr., Itasca, IL   60143-3201
(630) 285-1121 • fax: (630) 285-1315
e-mail: info@nsc.org
Web site: www.nsc.org

The National Safety Council is concerned with consumer safety in all phases of a product's use, including its safe and environmentally sound disposal. Through the organization's Electronic Product Recovery and Recycling Project (EPR2) it offers help to consumers who wish to recycle and reuse used electronics. Its publications include the *EPR2 Baseline Report: Recycling of Selected Electronic Products in the United States* and presentations from conferences on e-waste recycling.

Natural Resources Defense Council (NRDC)
40 W. Twentieth St., New York, NY   10011
(212) 727-2700 • fax: (212) 727-1773
e-mail: nrdcinfo@nrdc.org
Web site: www.nrdc.org

NRDC is dedicated to fighting for environmental protection through the use of science and law. It generally takes a hard line when dealing with issues of waste disposal. Its publications include "The Environmental Justice Movement" and "Burial Ground: Fear and Loathing at Yucca Mountain," both available on its Web site.

Resource Institute for Low Entropy Systems (RILES)
179 Boylston St., 4th Fl., Boston, MA   02130
(617) 524-7258 • fax: (617) 522-0690
e-mail: info@riles.org
Web site: www.riles.org

The Resource Institute for Low Entropy Systems is an independent nonprofit organization that works in partnership with communities in English- and Spanish-speaking countries to protect public health and the environment. RILES generally takes the view that industrial and municipal waste output must be strictly regulated and supports "non-depleting, non-wasting, non-polluting methods and technologies for sustainable development." Its publications include a book available in electronic form, *Toward Sustainable Sanitation*, as well as shorter articles under the title "Weekly Musings" available on its Web site.

Sierra Club
National Headquarters, San Francisco, CA   94105
(415) 977-5500 • fax: (415) 977-5799
e-mail: information@sierraclub.org.
Web site: www.sierraclub.org

The Sierra Club is the most prominent environmental group in the United States. The organization favors strict regulation of industry in order to protect the environment. It publishes a variety of books and reports, including official policies adopted by the club, such as "Municipal Solid Waste Management" and "Low-Level Radioactive Waste," which are available on its Web site.

Silicon Valley Toxics Coalition
760 N. First St., San Jose, CA   95112
(408) 287-6707 • fax: (408) 287-6771
e-mail: svtc@svtc.org
Web site: www.svtc.org

This group was founded in 1982 to tackle the issue of toxic waste from supposedly "clean" industries such as electronics and computer manufacturing. Since then it has expanded to global concerns over e-waste and promotes recycling, reduction, and reuse in the electronics industry. Its publications include *Challenging the Chip*, a history of activism against e-waste, and *Campus Report: System Error*, a guide for student-run campaigns against pollution by the electronics industry.

Union of Concerned Scientists

2 Brattle Square, Cambridge, MA 02238-9105
(617) 547-5552 • fax: (617) 864-9405
e-mail: erobinson@ucsusa.org
Web site: www.ucsusa.org

The Union of Concerned Scientists focuses strongly on safety issues concerning nuclear power. The group monitors the nuclear industry and generally opposes increased use of nuclear power, citing, among other problems, the lack of a secure disposal strategy for the radioactive waste generated. Its publications include short news articles such as "DOE Proposes Nuclear Waste Dumps for 11 Communities" as well as longer reports such as *Clean Energy*.

Waste to Energy Research and Technology Council

Earth Engineering Center, Columbia University
New York, NY
(212) 854-5213 • fax: (212) 854-9136
e-mail: earth@columbia.edu
Web site: www.seas.columbia.edu/earth/wtert

This academic organization promotes recovery and recycling of garbage and the use of solid waste for energy generation. As scientists and engineers, its members generally believe that waste problems can be solved to a great extent via technology. Its publications include *Energy from Waste, State-of-the-Art-Report, Statistics*, and *Renewable Fuels: A Critical Part of a Sustainable Energy Policy*. The group's publications are generally available on its Web site.

Bibliography of Books

Terry L. Anderson, ed.
You Have to Admit It's Getting Better: From Economic Prosperity to Environmental Quality. Stanford, CA: Hoover Institution Press, 2004.

Stewart Barr
Household Waste in Social Perspective: Values, Attitudes, Situation and Behaviour. Burlington, VT: Ashgate, 2002.

Pieter van Beukering
Recycling, International Trade, and the Environment: An Empirical Analysis. Boston: Kluwer Academic, 2001.

Harvey Blatt
America's Environmental Report Card: Are We Making the Grade? Cambridge, MA: MIT Press, 2005.

Michael Brower and Warren Leon
The Consumer's Guide to Effective Environmental Choices: Practical Advice from the Union of Concerned Scientists. New York: Three Rivers, 1999.

Melissa Checker
Polluted Promises: Environmental Racism and the Search for Justice in a Southern Town. New York: New York University Press, 2005.

William A. Cohen and Ryan Johnson, eds.
Filth: Dirt, Disgust, and Modern Life. Minneapolis: University of Minnesota Press, 2004.

Pat Costner and Greenpeace USA
The Burning Question: Chlorine and Dioxin: A Greenpeace Report. Washington, DC: Greenpeace USA, 1997.

Luis F. Diaz et al. *Solid Waste Management*. Paris: United Nations Environment Programme, 2005.

Christopher H. *The Promise and Peril of Environmen-*
Foreman *tal Justice*. Washington, DC: Brookings Institution, 1998.

Howard Frumkin, *Environmental Health: From Global to*
ed. *Local*. San Francisco: Jossey-Bass, 2005.

Don Fullerton *The Economics of Household Garbage*
and Thomas C. *and Recycling Behavior*. Northamp-
Kinnaman ton, MA: Elgar, 2002.

R.E. Gephart *Hanford: A Conversation About Nuclear Waste and Cleanup*. Columbus, OH: Battelle, 2003.

Kenneth N. *The Greening of Pentagon Brownfields:*
Hansen *Using Environmental Discourse to Redevelop Former Military Bases*. Lanham, MD: Lexington, 2004.

Lis Harris *Tilting at Mills: Green Dreams, Dirty Dealings, and the Corporate Squeeze*. Boston: Houghton Mifflin, 2003.

David Hosansky *The Environment: A to Z*. Washington, DC: CQ Press, 2001.

Gillian Klucas *Leadville: The Struggle to Revive an American Town*. Washington, DC: Island, 2004.

John Knechtel, ed. *Trash*. Cambridge, MA: MIT Press, 2007.

| Ruediger Kuehr and Eric Williams, eds. | *Computers and the Environment: Understanding and Managing Their Impacts.* Boston: Kluwer Academic, 2003. |

| Edward A. Laws | *Aquatic Pollution: An Introductory Text.* Hoboken, NJ: Wiley, 2000. |

| David A. Lochbaum | *Nuclear Waste Disposal Crisis.* Tulsa, OK: PennWell, 1996. |

| Christian V. Loeffe, ed. | *Conservation and Recycling of Resources: New Research.* New York: Nova Science, 2006. |

| Allison Macfarlane and Rodney C. Ewing | *Uncertainty Underground: Yucca Mountain and the Nation's High-Level Nuclear Waste.* Cambridge, MA: MIT Press, 2006. |

| Alexander D. Maples | *Sustainable Development: New Research.* New York: Nova Science, 2005. |

| Martin V. Melosi | *Garbage in the Cities: Refuse, Reform, and the Environment.* Pittsburgh: University of Pittsburgh Press, 2005. |

| Eric W. Mogren | *Warm Sands: Uranium Mill Tailings Policy in the Atomic West.* Albuquerque: University of New Mexico Press, 2002. |

| Garth Andrew Myers | *Disposable Cities: Garbage, Governance and Sustainable Development in Urban Africa.* Burlington, VT: Ashgate, 2005. |

Organisation for Economic Co-Operation and Development

Addressing the Economics of Waste. Paris: Organisation for Economic Co-Operation and Development, 2004.

David N. Pellow

Garbage Wars: The Struggle for Environmental Justice in Chicago. Cambridge, MA: MIT Press, 2002.

William L. Rathje and Cullen Murphy

Rubbish! The Archaeology of Garbage. Tucson: University of Arizona Press, 2001.

Lasse Ringius

Radioactive Waste Disposal at Sea: Public Ideas, Transnational Policy Entrepreneurs, and Environmental Regimes. Cambridge, MA: MIT Press, 2001.

John Scanlan

On Garbage. London: Reaktion, 2005.

Debra L. Strong

Recycling in America: A Reference Handbook. Santa Barbara, CA: ABC-CLIO, 1997.

Carl A. Zimring

Cash for Your Trash: Scrap Recycling in America. New Brunswick, NJ: Rutgers University Press, 2005.

Index